T0291991

Energy and the Missing Resource provides a nontechnical analysis of present and future energy resources and their potential development to meet future demand.

The prevailing impression in popular discussion of future energy supply is that a crisis will occur, sooner or later, owing to the exhaustion of present resources. In this informative and thought-provoking book, Professor Dostrovsky, who is a leading energy researcher, demonstrates that sufficient resources are available to meet all energy needs for the foreseeable future. However, this does not remove the threat of an energy-supply crisis. What is lacking – the missing resource – is the *knowledge* of how to use these resources in a practical and environmentally acceptable manner.

The author argues that long-term technical development will be necessary to ensure future energy sufficiency and that international cooperation on technical research, environmental impact, and energy use is needed now to prevent a succession of energy crises in the future.

The book avoids technical jargon, and mathematics is kept to a minimum. All those involved with energy in a technical, business, or governmental policy capacity will find this book essential and rewarding reading.

Energy and the missing resource

Energy and the missing resource

A view from the laboratory

I. DOSTROVSKY

Weizmann Institute of Science

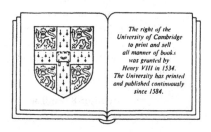

The right of the
University of Cambridge
to print and sell
all manner of books
was granted by
Henry VIII in 1534.
The University has printed
and published continuously
since 1584.

CAMBRIDGE UNIVERSITY PRESS

Cambridge

New York   New Rochelle   Melbourne   Sydney

CAMBRIDGE UNIVERSITY PRESS
Cambridge, New York, Melbourne, Madrid, Cape Town,
Singapore, São Paulo, Delhi, Mexico City

Cambridge University Press
The Edinburgh Building, Cambridge CB2 8RU, UK

Published in the United States of America by Cambridge University Press, New York

www.cambridge.org
Information on this title: www.cambridge.org/9780521319652

© Cambridge University Press 1988

First published 1988

A catalogue record for this publication is available from the British Library

Library of Congress Cataloguing in Publication Data
Dostrovsky, I.
Energy and the missing resource : a view from the laboratory /
I. Dostrovsky.
p.   cm.
Bibliography: p.
Includes index.
ISBN 0-521-26592-4   ISBN 0-521-31965-X (pbk.)
1. Power resources.   2. Power (Mechanics)   I. Title.
TJ163.2.D68   1988   88-14916
333.79–dc19   CIP

ISBN 978-0-521-26592-8 Hardback
ISBN 978-0-521-31965-2 Paperback

To the memory of
"Yoka"
Joseph Adar
recalling the long discussions about energy on land
and on the high seas

He that will not apply new remedies must expect new
evils; for time is the greatest innovator.

Francis Bacon

Contents

Figures and tables

Figures

Tables

Preface

The interest in energy issues waxes and wanes in keeping with the fluctuations in oil prices. The preoccupation with these short-term changes tends to obscure the long-range energy problems so crucial to mankind's future.

This book is devoted primarily to these longer-range problems. The discussion is necessarily oriented to the questions of energy resources and the technologies for their possible exploitation. The justification for dealing at the present time with issues that may seem to relate more to the next century than to the present one is the very long lead times involved in the development and deployment of new major technologies. This little-appreciated fact makes it essential to start action now if future energy crises are to be avoided.

The present time, with its respite from the energy crisis, is appropriate for a calm evaluation of the range of problems that face mankind in its search for inexhaustible and environmentally benign energy sources. It is hoped that this book may contribute somewhat to the education of the public and aid the informed debate on these issues.

The book is not a proselytizing one trying to "sell" a particular solution. Nor does it pretend to predict what the future may be. Rather it takes the approach of "what might happen if we do or do not do this or that."

I am grateful to my wife, Daphne, for her constant encouragement and support throughout the long evolution of this book and for her help in weeding out the trite. Thanks are also due to the many colleagues in several countries with whom the various issues raised in the book were discussed, in particular to (the late) J. Adar and to S. C. Cannon, F. Kreith, and I. Spiewack for reading the manuscript and making valuable comments. The views expressed in this book are solely my own and do not necessarily represent those of the institutions with which I am associated.

1

Introduction

For most of its evolution, mankind relied for its sources of energy on constantly replenished materials. When the use of fire was discovered, for the provision of heat and for the processing of food, the additional demand for energy was met by constantly renewed sources. Later still, when water and wind power were harnessed to the service of mankind, the new sources were also of a renewable nature. Thus, throughout this early phase of human development, the availability of the readily renewable sources of energy was a key constraint and affected the size and distribution of populations.

It is only relatively recently that mankind turned to the large-scale exploitation of new, and nonrenewable, sources of energy, and these sources have made possible the advance of the industrial revolution and the large increase in populations. It is clear that because of the finite magnitude of these nonrenewable sources, this phase of human development is transitory and sooner or later must come to an end. Mankind will again have to depend on renewable or practically infinite energy sources. On the long evolutionary time scale, therefore, the era of extensive dependence on the finite fossil-energy resources will appear as a brief episode.

For mankind to return willingly to the situation that prevailed a few thousands of years ago, with the present size and distribution of the population, is clearly unthinkable. It is necessary to examine, therefore, whether the knowledge and experience gained, and which still may be added to, in this era of plentiful and convenient energy sources, can be used to steer mankind back to a dependence on permanent energy sources. It is to this issue that this book is primarily addressed, and it is therefore a long-range review of the energy problem. How long depends on one's assessment of the lifetime of the existing major fossil-energy sources and other constraints to their use. Although the longer-range view is necessarily imprecise, it is freer from considerations arising out of temporary fluctuations in supply and demand.

At first sight it might be thought that long-term considerations are less important and certainly not urgent. But it is in the nature of things that time inexorably converts a long-range problem into an immediate and urgent one. More important is the fact that the times needed for the development of new technologies are also very extended and may be commensurate with the long-range problems that they have to solve.

We may distinguish three main constraints on the availability of energy: (1) quantitative bounds on the amount of a given resource at acceptable cost, (2) environmental limitations on the exploitation of the resource, and (3) lack of knowledge of how to use the resource.

Discussions of the limitations of energy supplies usually concentrate on the first constraint. The common approach to the assessment of the adequacy of a given resource is to make an estimate of its magnitude, under certain technological and economic assumptions, and to compare this estimate with some future demand forecasts according to some scenario. It is clear that this procedure is open to numerous biases, hidden or explicit, and to personal value judgment in all its steps. It is largely for this reason that the debate on energy issues is often acrimonious and inconclusive.

Although there is no doubt that the absolute exhaustion of a given resource sets a final limit on its use, in reality such an abrupt end is never reached. There is a long period during which the resource becomes scarcer and scarcer, its cost rises, and its pattern of use changes upon the introduction of substitutes. In addition, there may well be other constraints that come into play long before the total exhaustion of the resource limits its use.

In this book we shall assume that the shorter-range actions, such as energy conservation and increased efficiency of utilization, have been implemented to their full extent, and we shall not discuss them further (see, however, Section 6.2). We shall concentrate our attention on the information needed to bring in new and additional resources and on an understanding of their environmental and social limitations.

Several primary energy sources together can supply the needs of modern societies. Some are finite in extent (e.g., fossil fuels), some are practically unlimited (e.g., deuterium), and others are truly renewable (e.g., solar energy). Unfortunately, they are not all equally useful or interchangeable, and so it is not enough to deal with the total energy resources available; one must instead consider each source separately and study its limitations. Chapters 2 to 5 are devoted to such an analysis of each primary energy source. The analysis briefly reviews the origins of the resource, its estimated magnitude, the technologies used for its exploitation today, and its field of application. It ends with a discussion of the scientific and technological barriers to an increase in the reserves and an intensification of its use. Environmental and social constraints are also discussed where appropriate. These constraints are, of course, susceptible to change as a result of further research. But first we must define and clarify the terms and concepts used in this book.

1.1 Renewable and nonrenewable resources

The division of energy sources into renewable and nonrenewable categories is somewhat inaccurate – the real interest is whether a particular resource can provide energy for mankind in perpetuity or whether it is transient in nature. The term *perpetuity* needs some clarification for our discussion, for nothing is strictly perpetual. Even solar energy, commonly regarded as a perpetual source of energy, is the result of the "burning" (by a fusion process) of the nonrenewable hydrogen present in the core of the sun. The amount of hydrogen in the sun is estimated to last for a few billion years more, and it is apparent that such a time scale is considered as perpetual. Actually, much shorter time scales, of the order of a few million years or less, are also accepted as practically infinite as far the future of mankind is concerned. We may conclude, therefore, that any source of energy having the potential of lasting a million or so years can be regarded as a perpetual source as far as the present discussion is concerned, whether that source is renewable or not.

An energy source can be perpetual if it is renewed continuously, and on a short time scale, from another source of a lifetime much longer than a few million years, or if it is present in an amount sufficient for a few million years. Examples of the first kind are all those sources that depend basically on solar energy for their renewal, such as biomass, wind, and hydropower. An example of the second type is nuclear energy.

We shall discuss some of these examples further to clarify the various considerations. The amount of oil available to mankind certainly falls far short of supplying the needs for millions of years. And although it is probable that oil is still being formed today from buried biomass (i.e., from solar energy) in some geological structures, the rate of this process is so slow compared with exploitation that it is entirely insignificant. Oil can therefore be regarded as a truly nonrenewable resource. Similar considerations can be applied to coal and gas.

Although renewable resources are, by definition, inexhaustible, the potential of some of them is limited, because it depends on their rate of renewal. Obviously, one cannot use such a resource at a rate faster than its rate of renewal. For example, wood as a source of energy is certainly renewable on a time scale of decades. However, its total potential is limited, and we have already witnessed crises when demand for this source of energy exceeded the rate of growth of trees. Hydropower is also renewable on a time scale of years, but the total potential is limited by the rate of precipitation. Actually, in reservoir-based hydropower the potential is slowly decreasing as the result of silting (see Section 5.3).

But what about uranium? It is certainly not being formed on earth today and is therefore strictly nonrenewable. However, the magnitude of the resource is such that, if we include extraction from the oceans and assume breeding technology (both feasible, if uneconomical), it can supply the total present electricity demand for over a million years (see Section 4.3). We can therefore regard it as a perpetual, although nonrenewable, resource. An even more extreme example of this kind is the isotope deuterium, which is present in a proportion of 1/6700 of all the hydrogen atoms found in the ocean and its quantity is therefore enormous. It is not produced today, and yet if deuterium-burning fusion ever becomes technologically and economically feasible, it will suffice for human needs for many millions of years. So although uranium and deuterium (and thorium) are not strictly renewable, they can be considered, on the time scale of human civilization, as perpetual.

1.2 Reserves and resources

It is only recently that mankind awoke to the fact that many natural resources are finite and face exhaustion in a future that is not too far away to be perceived. When the world's population was relatively small and growing slowly and there were still large unexplored areas of the earth, most people felt that natural resources were unlimited. The exhaustion of a local source of valuable material (gold, copper, etc.) could always be more than compensated for by fresh discoveries in newly explored lands. Indeed, one of the motivating forces for exploration was the search for new resources.

In connection with energy, the realization that a practical limit to the resources exists and that it may be not too far in the future is of relatively recent origin, and such awareness is not universal. There is still the feeling, based on historical development, that when the need arises fresh discoveries will bring in new and larger resources. To support this view, people point out past predictions of the imminent exhaustion of oil and coal. It is also often hinted that the various estimates are not always devoid of ulterior motives and must therefore be suspect.

Research in the last half century has enormously increased our understanding of the origins of the present main energy sources, the fossil fuels (i.e., coal, oil, gas), the mechanisms by which they are formed, and the likely geological formations where they may be found. At the same time, most of the earth's surface has been explored to some extent, and its major geological features are known. This knowledge enables one to make rough global estimates of the various resources, particularly oil.

In dealing with finite natural resources, we encounter a broad range of deposits that differ not only with respect to the grade of the material but also in the extent of our knowledge about them and the difficulties of their practical extraction and utilization. The quantity of some deposits may be known with great confidence, but the amounts of others may be only guesses; some may be easier and cheaper to extract than others. Yet in other cases there may be no known technology of recovery. Because of all this, no two deposits are exactly alike in all respects.

Several definitions are in use to distinguish between the various types. We will apply the term *reserve* to those quantities of the material of interest (oil, coal, gas, or uranium) that have been proved by measurement and that may be economically recovered by known technology. We will use the term *resources* to describe those additional quantities of the material assumed to exist on the basis of superficial exploration or from general geological knowledge, and also material that, although proved to be present, cannot be economically won by present technologies. These definitions leave much to be desired as far as intellectual rigor goes, but they do represent the true state of affairs and the uncertainties involved.

Reserves may, and do, change as a function of time. This can happen either as a result of new measurements (e.g., drilling and analysis), because of the development of new technologies, or simply because of the change in economic circumstances. The last factor provides a damping negative-feedback mechanism. As prices rise, some resources become reserves worth exploiting, and then the additional material that becomes available on the market slows down the rate of price increase. Reserves always decrease as the result of exploitation.

Resources can change in both directions by further exploration and in general by increased understanding of the geological and chemical processes on earth. The rate of exploration itself is a function of the economic conditions and, therefore, changes with time.

A good deal of effort has been devoted to estimating resources of the various fuel forms, and the effort continues. Obviously, it is very important for present and future planners to know how long fuels will last. Interest in the subject of the magnitude of the resources is not new, but it has reached high pitch with the general realization that they are finite and that the lifetime of some of them may be such as to require some action now rather than in the distant future. We shall discuss these issues in greater detail in connection with individual fuels in later chapters.

1.3 Units

Before proceeding with more quantitative discussions of reserves and resources, we must describe the units used to measure them. Historically the energy field developed as a number of loosely related disciplines, and it is therefore not surprising that each discipline adopted its own system of units and standards. In addition, the English-speaking world uses units of weights and measures different from the rest of the world. Over long periods of time various international usages were adopted and changed again. The result of all this is a proliferation of units by which energy and power are measured and reported. Thus British thermal units (BTUs) are used by the English-speaking world, and calories by the rest, with joules now being introduced to replace both as units of energy. Kilowatt-hour is used by electrical utilities, BTUs (or calories) per cubic foot (or cubic meter) are used by the gas industry, and millions of electron volts (MeV) are used by nuclear scientists.

Naturally the coal industry uses tonnes of coal equivalent (TCE) to measure its energy output, whereas the oil industry uses tonnes of oil equivalent (TOE) or barrels of oil. To increase the confusion, three different kinds of ton are used: the long ton (2240 lb), the short ton

(2000 lb), and the metric ton (tonne = 1000 kg). The situation in units of power is not much better. Thus, we find kilowatts (or megawatts) used by the utilities, and horsepower used in general industry.

Unusual multipliers of fundamental units are also used. Thus, for example, we find the therm (10^5 BTU) and quads (10^{15} BTU). This multiplicity of units, some of which are ill defined, not only confuses the public and sometimes prevents, or misleads, the public from understanding the issues discussed, but it also hinders communication among practitioners in the energy field. Another effect of the multiplicity of measures is that it tends to conceal the essential unity of the energy field.

The interest in global aspects of energy also led to the use of very large numbers, of whatever units, to describe amounts produced or consumed. The names of the exponents used to describe these large numbers in scientific notation (in powers of 10) are also not generally familiar. All these difficulties are compounded by the often hazy notions of what is involved in the interconversion of the forms of energy. For convenience, Appendix A at the back of the book lists conversion factors and definitions of some of the most useful units used and the prefixes of the large numbers involved in the energy field. Such a list has become a necessary feature in any book on energy. We shall use, as far as possible, the International System of Units (SI). However, these are not commonly used in the energy field, and they lack the intuitive impact of the everyday units, so we shall employ the latter when necessary.

Before leaving this subject, we discuss the very commonly used units tonnes of coal equivalent or tonnes of oil equivalent. Coals from different sources vary considerably in their calorific value (i.e., the amount of heat produced on their combustion). Thus a low-volatile bituminous coal may assay as high as 36.3 GJ/tonne, and a high-volatile bituminous coal may assay 28.8 GJ/tonne. Typical lignites assay at 17.7 GJ/ton. By convention the United Nations defined 1 metric ton of coal to be equivalent to 7000 million calories (29.3 GJ), whereas in the United Kingdom the conversion is 26.4 GJ. In using these conversions, one must ascertain what kind of ton is referred to. In the oil business also there is a variation in the density and calorific value of material from various sources. Thus a tonne of oil may produce between 42 and 45 GJ, depending on source. By convention again, 1 tonne of oil equivalent (TOE) is now taken as producing 44 GJ of heat. One must be careful when using assay numbers to ascertain whether high or low calorific values are being reported. These values differ according to whether or not the latent heat of evaporation of water (a combustion product) is considered. The two numbers may differ by as much as 10% for natural gas, 5% for oil, and only 2% for coal; the higher value includes the latent heat of water.

The reader is perfectly justified if he or she is confused and mystified by this state of affairs. It is one of the hazards one must face in trying to understand the operation of the energy economy.

A particular problem arises when some common basis is required to discuss resource bases of electricity generated by nuclear reactors or hydroelectric schemes together with that produced from fossil fuels. The usual practice, adopted in this book, is to express nuclear energy or hydropower for such purposes as the equivalent of the amount of fossil fuel that would be used in a thermal station to produce the same amount of electricity. Clearly, this is not a very rigorous definition, because it depends strongly on the efficiency of conversion of the thermal station (its "heat rate"). Here again the convention (as used by the U.S. Department of Energy) is to assume that 10,435 BTU (11.01 MJ) of heat are needed to produce a kilowatt-hour of electricity. Using this rate, we have the following identities:

$$1 \text{ million kWh(e)} = 11,009 \text{ GJ} = 10,435 \text{ million BTU}$$
$$= 255 \text{ metric tons oil}$$

The reader must be cautioned to use these conversion factors only in the limited context of comparing electricity production from nonfossil sources with that generated from fossil fuels. These factors differ from the normal conversion factors presented in Appendix A. Thus, 1 kWh of electricity put through a resistance heater will still generate only 3.6 MJ of heat (and not 11.01 MJ).

1.4 Interconversions

We have noted that energy can be converted from one form to another. Some energy conversion processes have been practiced throughout history. Thus, the conversion of wind power to mechanical energy (for pumping water) is recorded as early as the seventh century A.D. The use of water power to drive machinery is also an ancient art. However, the conversion of heat to mechanical energy is a much later development. This may reflect the fundamental difference between the processes. On the whole, fuels were used directly to generate heat where needed without any processing. Thus, wood was burnt to provide comfort and for cooking probably from the earliest days of the discovery of fire; later, coal was burnt in a similar way. These processes represent the conversion of chemical energy (combustion) to heat. Some processing of the fuel for special purposes was practiced later still, such as conversion of wood to charcoal or of coal to coke, but these are not energy conversion processes. Similarly, the refining of oil and its splitting into various fractions is not an energy conversion process. One can then say that until the invention of the steam engine wind and water power were the only examples of energy conversion practiced, and even then the conversion was from mechanical energy (movement of air or water) to mechanical energy.

No energy is lost in an interconversion process, although not all the initial energy ends up in the particular form we desire. The efficiency of conversion for many processes can in theory be 100%. In practice, however, because of friction and other losses, such efficiency can never be attained. The part of the energy not converted to the desired form usually appears as heat. We shall illustrate the point by a few examples. Processes that involve the interconversion of mechanical and electrical energies have a theoretical efficiency of 100%, and in practice can reach values close to this. Thus the efficiency of a modern electric motor (converting electrical energy to mechanical energy) is about 97%. Similarly, the efficiency of a modern electric generator (converting mechanical energy to electricity) is about the same. The loss (a small percentage) compared with theoretical efficiency appears as heat in the bearings, in the windings (ohmic losses), and in the metal core (eddy currents).

When we turn to consider processes in which heat is converted to other forms of energy, we find the situation very different. It is theoretically impossible to convert quantitatively all the energy present as heat to other forms, and in practice the theoretical limit is never reached. Some, and often most, of the energy must be degraded in the conversion process to lower-temperature heat. The science of thermodynamics was developed as a result of interest in the processes of energy interconversions, particularly those involving heat.

To understand the arguments presented in this book, the reader need only have some familiarity with the two basic laws of thermodynamics. We have already mentioned the first law, without naming it specifically, when we stated that energy is not lost in any conversion process. A corollary of this is that energy cannot appear out of nowhere. In a more formal vein, the first law of thermodynamics states that the total sum of all kinds of energy in a closed (isolated) system is a constant. Within such systems energy may be converted from one form to another in many different processes, but the sum total of all forms of energy remains unchanged.

The other basic law relates to the cumulative experience that the spontaneous direction of heat flow is from a hot source to a cooler sink, and not the other way around. The second law will be disproved when someone puts a kettle of water on a heater and later finds a lump of ice in the kettle. There are many equivalent ways of defining the second law. Because our primary interest at the moment is in the field of energy conversion processes, particularly in the conversion of thermal energy to other forms, most frequently mechanical, we shall use the following statement (paraphrasing Thomson (Lord Kelvin) 1851): "It is impossible to convert, in a cyclic manner, heat obtained from a reservoir to another form of energy without rejecting some heat to another reservoir at a lower temperature." In other words, it is only possible to convert thermal energy to some other form continuously if a reservoir at a higher temperature than a heat sink is available. This statement is equivalent to saying that a perpetual motion of the second kind (i.e., a device that will extract energy from one reservoir and convert all of it to another form) is impossible.

It can be readily shown, as done by Sadi Carnot in 1824, that the maximum efficiency of conversion of an ideal lossless heat engine depends only on the temperatures of the reservoir and the heat sink. The relation is very simple mathematically: The maximum theoretical efficiency of conversion of heat to other forms of energy equals the the difference in temperatures between the hot (T_h) and cold (T_c) reservoirs divided by the temperature (on the absolute scale) of the hot reservoir. Thus,

$$\text{efficiency} = \frac{T_h - T_c}{T_h}$$

This simple formula determines the entire technological development of the energy conversion industry. Note that it is desirable to have as low a sink temperature and as high a source temperature as practicable (in spite of the presence of the T_h term in the denominator). In practice, any thermal plant will use as a heat sink the coldest large body of water available, usually the ocean or a river. Failing that, a cooling tower will be used to reject the heat to the atmosphere. The highest temperature is usually determined by the properties of the materials from which the boilers and turbines can be made. The technological developments of the power industry have been directed to reaching higher and higher operating temperatures. The limit for steam turbines at present is about 550°C.

In practice, it is impossible to achieve the maximum Carnot efficiency, and it will never be possible. The Carnot efficiency is calculated for ideal conditions, which do not exist in the real world. The challenge to engineers, however, has been to try and approach the Carnot efficiency as closely as possible. Much ingenuity has been devoted to this problem, and the present record in large thermal plants is probably an approach to within 80% of the maximum Carnot efficiency.

We give some examples. For a modern thermal power station the upper temperature is about 550°C, and the cooling water is around 30°C. These temperatures would lead to a Carnot efficiency of 63%. The actual efficiency is about 40% – that is, about 63% of Carnot. It is possible to approach more closely the theoretical efficiency, but the cost of the plant will be greatly increased. The values actually chosen are the result of careful optimization for minimum electricity costs. As the relative costs of fuel and capital change, so will the optimization point. The simple efficiency formula should enable the reader to be on guard against irresponsible claims of efficiencies of new inventions. Any claim for a thermal conversion process with efficiencies higher than or even approaching Carnot should not be taken seriously.

2
Fossil fuels

Fossil fuels are materials that have originated as living matter and have been altered by chemical processes over geological times to yield substances containing a high proportion of carbon. Fossil fuels include oil, natural gas, coals, and materials such as oil shale, tar sands, and peat. The ultimate source of the energy released upon the combustion of all fossil fuels is the solar radiation trapped many millions of years ago by living organisms.

2.1 Oil

Crude oil is a complex mixture of hydrocarbons and some organic compounds of sulfur, nitrogen, and oxygen. The molecular complexity extends from simple short chains of the light hydrocarbons, containing only a few carbon atoms, to complex structures of the asphaltenes, containing over 40 carbon atoms. These substances are formed as the breakdown products of plant organisms, mainly of marine origin, that become incorporated in sediments and are then subjected to heat under high pressures over long periods of time. The conditions under which the organic material can be converted to petroleum are rather critical, and so only a small fraction of the total biomass produced in the oceans and lakes or brought into them by rivers is ultimately transformed to crude oil. Regions of high biological productivity, such as zones of upwelling and inshore areas, will obviously be potential sources of sediments rich in organic material. However, this is not enough to ensure the formation of potential oil sources; the precipitated organic matter must escape oxidation by oxygen dissolved in the water. Where stagnant conditions exist, accumulation of sediments rich in organic debris may be formed. Such sediments, when compacted by extensive pressure of accumulated material, become rocks, *source rocks* as they are called in the petroleum industry, in which oil may be formed. The organic content of deep ocean sediments is only about 0.1% but may rise in suitable locations (e.g., the Black Sea) to 15%.

As the depth of burial of a sediment increases, so do the pressure to which it is subjected and its temperature. The latter is a result of the geothermal gradient, which in most normal areas is about 3°C/100 m. The chemical transformation to hydrocarbons of the organic material in the rocks begins at about 65°C, at which temperature the process may require about 400 million years for completion. As in all chemical reactions, an increase in temperature causes an increase in the rate of reaction, so at about 140°C the conversion to oil in the source rocks takes about 10 million years. At temperatures appreciably above 140°C the decomposition of the organic matter gives primarily gaseous products. We can see therefore that neither a young sediment in a region of low thermal gradient nor an old deposit subjected to high temperatures for a long time will be likely to yield oil. Any oil formed long ago under the latter conditions either escaped or was decomposed.

The crude oil formed in the source rock by the decomposition and reprocessing of the organic matter is found as a fine dispersion in the matrix and cannot be exploited directly. Over geological time, however, various processes can cause the crude oil in the source rocks to migrate to and along porous strata. This migration can continue until the oil reaches the surface and is lost, or until it reaches a trap in which it accumulates (the reservoir). Traps are geological formations that are sufficiently porous to store appreciable amounts of fluids but from which the further migration of oil is prevented by an overlying impermeable layer (often an anticlinal structure) or by a discontinuity (e.g., a fault) that blocks an exit from the porous layer. The art of oil prospecting consists in identifying suitable trapping structures in regions where source rocks meeting the stringent criteria just outlined are present.

As geological knowledge about the world's past sedimentary basins increases, it becomes possible to estimate with more confidence the probability of existence of oil resources in unexplored regions. Estimates of the maximal future potential oil resources are based on such considerations. When we come to the determination of actual reserves, such general considerations are not sufficient. A drilling program has to be undertaken to prove the existence of the reservoir, its magnitude, the quality of the oil and other factors that may determine whether production is economically worthwhile under existing conditions. A battery of techniques is at the disposal of the modern prospector. These include seismic surveys, gravity surveys, electrical and magnetic measurements, radioactivity measurements, and geochemical surveys. In spite of all these technologies the rate of success in exploratory drillings, as measured by the proportion of "dry" holes or by the increasing footage drilled per discovery, is low and is getting lower. This illustrates the increasing difficulty of bringing in new oil sources.

Crude oil occurs in grades that differ in composition. The grades are characterized by terms such as *light* or *heavy* according to the proportion of the more volatile hydrocarbons present. The grades also differ in market value, ease of production, and ultimate recovery fraction.

Crude oils containing a high proportion of the volatile hydrocarbons of small molecular weight (chains of 5 to 10 carbon atoms) are more valuable because more light motor fuel can be recovered from them by simple fractional distillation. They are also less viscous and therefore flow more readily. This makes their production easier and their recovery more complete. The heavier crudes contain only a small proportion of the volatile hydrocarbons and consequently produce less light motor fuel upon simple fractionation. Special techniques, involving extra costs, are necessary in order to increase the proportion of gasoline in the refining process. The recovery of the heavy crudes is also more expensive because of their slow pumping rate. Unless special measures are adopted, their ultimate recovery is also low; it may be as low as 5% of the oil present in the formation.

It is common to distinguish three stages in the recovery of crude oil, each more costly than the preceding one. In the first, or primary, production stage the oil is recovered either by its own pressure, which drives it to the surface, or by simple pumping. When this step is no longer adequate to give the desired production rate, natural gas may be injected into the formation to increase the pressure above the oil, or water may be injected below the oil-bearing strata to flush it out. These techniques are known as *secondary recovery procedures*. They may add 5% to 10% to the recovery of the oil present. Further recovery may be obtained, particularly of heavy crudes, by tertiary procedures known as *enhanced oil recovery systems*. (For a review of this subject see Kuuskraa et al. 1983). The most common enhanced recovery method involves the injection of steam into the formation to heat the oil and thereby make it flow more readily. Impressive results have been obtained by this technique, and 30% to 70%

Table 2.1. *Estimate of recoverable resources of conventional oil (in exajoules)*

	Cumulative production	Remaining reserves	Most probable undiscovered	Ultimately recoverable
North America	893	396	1,025	2,314
South America	295	216	207	718
Europe (less USSR)	70	167	128	365
USSR	427	440	673	1,540
Africa	202	326	290	818
Middle East	779	2,790	788	4,357
Asia/Oceania	132	216	365	713
World	2,798	4,551	3,476	10,825

Source: Masters, Root, and Dietzman (1983).

of the remaining oil in place can be recovered. Another, more recent, method involves the injection of carbon dioxide under high pressure to act as a supercritical solvent for oil. Recoveries of 10% to 50% of the remaining oil have been achieved. Other procedures that may be used in tertiary recoveries involve injecting surface-active agents and various chemicals to release the oil adhering to the rock particles.

The average, worldwide, recovery fraction today is 34% of the oil in place in the formation. As the price of oil rises, greater use will probably be made of tertiary recovery techniques, and fields marginal because of their size or the quality of their oil will be brought into production. New techniques for enhanced oil recovery will probably also be developed. In this way the known reserves will be increased even without new discoveries.

In the last 20 years many estimates of the magnitude of the crude oil resource have been made; they are all in the range of 8000 to 15,000 EJ. Data released in the last few years, mainly by the oil industry, put the proven reserves at around 4000 EJ (90 GTOE). The various tabulations of known reserves or assumed additional resources are, however, based on recovery fractions achieved by the industry today. In Table 2.1 we present one of the latest estimates of the expected amount of conventional oil – oil that is recoverable by present technology. Note that according to the data of Table 2.1 just over a quarter of the resource

Table 2.2. *Estimates of heavy oil and tar sands in place (in exajoules)*

Venezuela (Orinoco)	4,400
Canada (Athabasca)	3,784
(Other)	1,848
USSR (Olenek)	3,784
United States	176
World	13,992

Source: Meyer et al. (1984).

ultimately recoverable by present technologies has already been used up. There remain about 8000 EJ of conventional oil or the equivalent of 182 GTOE.

2.1.1 Unconventional oil

In addition to conventional oil, there are in nature large deposits of much heavier oils that flow with great difficulty (API index less than 10) or do not flow at all under normal conditions (tar sands) and are therefore not recoverable by the usual technologies. Data on unconventional sources of oil are relatively meager because of the limited commercial interest in these materials at the present time. In Table 2.2 we list the known major deposits of these materials.

It will be seen from the data of Tables 2.1 and 2.2 that the heavy crude and tar sand resources (318 GTOE or 14,000 EJ) are close to those of the conventional oil remaining (182 GTOE or 8000 EJ) if one allows only 50% recovery for the former. These unconventional oil resources could undoubtedly be recovered, but at a cost higher than that of the conventional oil. The main experience in the recovery of oil from such sources is probably found in Canada. The huge deposits of tar sands found in Alberta have attracted attention since the end of the last century, but only in 1964 was the construction of an extraction plant begun. Tar sands contain between 8% and 12% of bitumen in a sandy matrix. There are several possible approaches to the recovery of crude oil from this material. If the formation is no deeper than about 50 m, open-cast mining can be employed and the solid mass sent (by conveyer belts) to the extraction plant nearby where the bitumen is separated from the sand by a hot-water treatment followed by decantation of the oily material that floats to the top. When the bitumen thus obtained is heated, it decomposes to yield gases, liquids, and a solid coke residue. The liquid fraction is taken, as synthetic crude oil, to conventional refineries for further processing. The other fractions are used in the process: The coke is burnt to provide steam, and the gases are used to make hydrogen. This is the process used in the Athabasca tar sand field to produce 50,000 bpd (barrels per day) of synthetic crude oil (1983). The properties of the tar sands and the severe climatic conditions in the regions where it is mined make this operation a test of technological and human endurance.

If the tar sand formation is deep below the surface, as is the case for 90% of the Canadian deposits, in situ recovery procedures must be used. One such procedure is very similar to that used in tertiary heavy-crude recovery. Steam at high temperature and high pressure is injected into the formation to heat it up. The hot bitumen is then pumped out and processed as described before. Another method is to use a combination of steam injection and in situ combustion to generate the heat required to increase the fluidity of the bitumen. Recoveries of bitumen range from over 90% for the open-cast mining to about 50% for the in situ procedures. The cost in 1978 was about $14 per barrel, which was higher than that of conventional crude at that time. Here again it is likely that as the price of oil rises in the future, these unconventional sources will become economically attractive and thus add considerably to the reserves.

2.2 Oil shale

Another possible source of unconventional and more expensive oil is oil shale. The term *oil shale* describes rather loosely all types of sedimentary rock containing an appreciable amount (over 3.5%) of organic matter. Oil shales probably approximate closely the source rocks of the oil genesis. They represent those sediments that have escaped so far the conditions (of temperature and time) that lead to alteration of the organic matter in them (known as kerogens) to form petroleum. These conditions may be induced artificially, by heating the shale in retorts, for example, and thereby recovering a crude oil from the material. Although it is likely that there exist very large resources of oil shale in the world, there is no precise estimate of the magnitude of these resources. In this century there has been very little interest in this energy resource, and what data are available are reported in numerous ways and based on

Table 2.3. *Major oil shale deposits in
the world (in exajoules of
recoverable oil)*

United States	12,239
Brazil	4900
USSR	690
Zaire	616
Canada	269
China	170
Rest of world	288
World	19,171

Source: Burger (1973); Matveev (1975).

different assumptions of what is included as a resource and what the
assumed recovery rate will be. In Table 2.3 we list some of the most
important known deposits. The potential of this energy resource (about
19,000 EJ) is greater even than that of unconventional oils.

A practical process for the recovery of crude oil from oil shale may
entail mining or quarrying the material, crushing it, and heating it in
retorts or other suitable reactors alone or in a stream of steam or hydrogen. It is estimated that over half the energy content of the shale will be
expended in these operations. Nevertheless, oil shale is an important
possible source of oil because much exists in parts of the world.

It is a common impression today that oil shale is a novel source that
may be of some use in the distant future. Actually, oil shale was the first
source of mineral oil that mankind used. The extraction of shale oil
preceded the discovery of petroleum by several decades. As early as
1838 there was a shale oil industry in France; in 1848 there was one in
Scotland. These industries actually survived well into this century, until
they were finally displaced by cheap oil in 1957 and 1963, respectively.
Somewhat later, shale oil industries were started in Germany, Estonia,
Spain, China, and Sweden. These industries are still operating, and
some are being enlarged. In China 330,000 tons of shale oil were produced in 1982, and in Estonia 680,000 tons were recovered in 1980.

Oil shale can be burnt directly in special boilers to generate steam and
electricity. More than half of the oil shale being mined in the USSR is
burnt directly at the pithead to generate electricity.

In the United States the development of a shale oil industry is still in its embryonic stages, but a 500,000-tons/yr shale oil demonstration plant has recently begun operating. In general, the cost of crude shale oil is higher than that of conventional oil. But, here again, the situation will probably change, and large-scale oil shale processing may become economically competitive when the price of conventional oil rises. One can therefore view the development of oil shale use as a process, begun in the last century, that after an interruption of 150 years or so by the incursion of cheap conventional oil could resume for several hundreds of years in the future (see, however, the discussion in Chapter 7).

The exploitation of oil shales and unconventional oil sources by surface mining entails a much more severe environmental impact than the production of conventional oil. The large open-cast mining operations and the high proportion of waste material produced are environmentally objectionable and may limit the extent to which these sources will be used. Because of this and because most of the oil shale deposits are below the depth of economical open-cast mining methods, great interest is shown in processes of in situ recovery. An added incentive is the fact that over 60% of the cost of recovery of shale oil by surface-based methods is due to the mining, crushing, transporting, retorting, and final disposition of the wastes. An effective in situ procedure avoids all these steps. Essentially, the method involves heating the oil shale in its formation to its decomposition temperature (around 500°C) by hot gases, which usually are obtained by the combustion of a part of the deposit. Since oil shales are dense nonporous rocks, the procedure can only work if the material is first fractured (or "rubblized") to provide easy passage of the heating gases and drainage paths for the oil that is formed. This oil is collected in sumps at the bottom of the formation, from where it is pumped to the surface. Several experimental in situ recoveries of shale oil were performed, but no commercial plant exists. The successful development of in situ recovery technologies is essential, if anywhere near the full potential of the oil shale and tar sands resources is to be realized.

2.2.1 Crude oil processing

Before the crude oil obtained from any source (conventional or unconventional or shale) can be used, it must be processed to yield the desired commercial products. The crude oil is transported from the wells or recovery plants (for unconventional oil) to the refineries by pipeline or tanker. The complexity of the refining process depends on the quality of the crude oil and the desired product mix. The basic operation is the fractional distillation of the crude oil to separate the various components according to their boiling points. For high-grade crude oils this operation is almost all that is required to give a range of salable products. Thus, the gases, consisting of the lowest hydrocarbons (those with chains containing 1 to 4 carbon atoms), are used either as a source material for the chemical industry or bottled and sold for domestic use as liquefied petroleum gases (LPG). The next fraction, consisting mainly of hydrocarbons with 5 to 10 carbon atoms, forms the basis of light motor fuels. Higher fractions provide kerosene, diesel fuel, grades of industrial fuel oils for steam raising and electric power generation, lubricating oils, and asphalts. During refining, the materials are purified from noxious substances, mainly sulfur-bearing compounds, and modified by additives to yield the specified end products for sale.

In general, however, the range of materials obtainable by a straight fractional distillation of crude oil does not correspond to the market demand for the various products. It often happens, particularly with heavier crude oils, that an excess of the high-boiling fractions and residue is obtained. To produce the range of materials desired, additional processes must be employed to convert some of the higher-boiling fractions into the valuable lower-boiling fuels (such as gasoline and kerosene), and these add to the cost of the operation. Considerable ingenuity is therefore applied to the choice of the crude oils processed and to the interconversion of the fractions by a series of chemical reactions. Consequently, a modern refinery is a very sophisticated and complex chemical factory.

2.3 Demand for oil

Of all the primary sources of energy, oil is the most versatile in meeting the whole range of demands. Thus, oil products are an important raw material for the petrochemical industry, and petroleum products are crucial to transportation in all its forms. Much of the industrial process heat requirement and domestic and commercial needs are satisfied by oil. And electric power generation in many parts of the world is heavily dependent on oil. In Table 2.4 we show a breakdown of the use of oil in the world by the major sectors.

Oil is so versatile not only because many diverse end-use products can be readily obtained from one raw material, but also because petroleum

Table 2.4. *Fraction of oil used in various sectors (OECD countries,*
1980)

Transport	Residential/ commercial	Industry	Electricity generation
0.43	0.20	0.26	0.11

Source: World Energy Conference (1983).

products are easy to transport and use. It is unlikely that another single
primary energy source will be found to replace oil in all its many uses.
The plentiful availability of oil in the last 100 years has made an indelible
impact on the pattern of development of civilization and the human way
of life. To maintain this way of life in the face of the changing sources of
primary energy is a major technological challenge. And to adapt our
present pattern of civilization to the new sources of energy is a major
social problem.

Significant variations exist in the pattern of use among countries at
various stages of development and in different regions of the world. If the
proportion of oil use in different sectors of the economy remained constant, the total demand for oil would follow the changes in the total world
energy requirements and could then be estimated somewhat more readily
(see Section 6.1). However, in some of its uses, oil can be replaced by
other sources of primary energy, so when the price of oil rises relative to
these sources, it will be replaced by the cheaper resource. This phenomenon is taking place in electric power production. In 1973, 25% of the
electricity in OECD countries was generated with oil as fuel; the proportion dropped to 11% in 1985, and is expected to reach 8% by 1990 (World
Energy Conference 1983). On the other hand, in some sectors, for example transportation and chemical feed stocks, oil is at present irreplace-
able. Therefore, the demand for oil will closely follow growth of these
sectors, at least until new technologies are developed.

2.4 Natural gas

Natural gas consists essentially of methane diluted by some other light
hydrocarbons and contaminated at times by carbon dioxide and hydrogen sulfide. These diluents or noxious gases must be removed before the

methane is shipped to consumers. Beyond this relatively simple operation, the raw material requires little processing.

Natural gas is formed as one of the products during the alteration of
organic matter contained in sediments under the influence of heat. The
process was described in connection with the genesis of oil (see Section
2.1). Recall that, beyond a fairly narrow temperature region, the main
product of the decomposition of the organic material is methane. This
material, being a gas, is very mobile and diffuses away from its point of
origin until it either escapes to the atmosphere or is trapped in a suitable
formation. Because the geological structures capable of trapping oil are
also effective in trapping gas, the two materials are often associated. For
the same reason the exploration for oil usually uncovers the gas, if
present. The proportions of the two, however, will depend very much on
the conditions to which the organic matter in the sediments was subjected. They can vary from all gas and no oil in the reservoir formation
to almost all oil in the heavy-crude deposits.

Methane is also formed in the processes that convert terrestrial plant
material to coal. The presence of methane in coal mines, the dreaded fire
damp, is perhaps the greatest hazard of the mining industry. But because
coal deposits are not necessarily associated with special trapping formations, the close association of the two products is not always found.

In recent years another, nonbiogenic, theory of natural gas formation
has been proposed (Gold 1985; Gold and Soter 1980, 1982). Methane is
present in space and is a component in the assembly of the solar system.
The massive outer planets Jupiter, Saturn, Uranus, and Neptune contain
large amounts of methane and other hydrocarbons in their atmospheres.
It is therefore reasonable to assume that methane was also a component
of the early earth. It has been suggested that some of this primordial
methane is still present and is seeping out slowly, so a fraction of it may
become trapped in suitable geological structures. If this postulate is true,
then methane need not be confined to sedimentary regions, but perhaps
can be found, even in larger amounts, in entirely new provinces. So far,
no clear evidence in support of this hypothesis has been accumulated, in
spite of a considerable amount of work. A deep drilling project in the
Siijan circle (an ancient asteroid impact crater) in Sweden has failed to
provide a proof of the abiogenic hypothesis.

For many years the discovery of a well yielding only gas was considered
a failure, and there was little incentive to search specifically for such wells.
For this reason the estimates of world reserves and resources were very

Table 2.5. *Major sources of natural gas*

	Proved recoverable reserves		Estimated additional recoverable reserves
	(Tm³)	(EJ)	(EJ)
USSR	35.0	1,211	4,152
Iran	11.0	433	236
United States	5.8	214	917
Algeria	3.1	117	261
Qatar	3.1	117	317
Canada	2.6	100	45
Mexico	2.2	86	191
Saudi Arabia	2.1	76	45
Venezuela	1.7	74	83
Netherlands	1.5	48	19
Norway	1.0	42	63
World	85.5	3,200	7,040

Source: World Energy Conference (1986).

patchy and unreliable. But in the last couple of decades, with the development of technologies for the transportation of natural gas across continents and oceans, the situation changed drastically. Major pipelines now crisscross continents, delivering natural gas to industrial and domestic consumers through complex secondary and tertiary networks.

The problem of supplying gas across oceans was solved by the development of low-temperature gas liquefication technology. Methane condenses to a liquid at −164°C at atmospheric pressure. The liquid can be stored in insulated containers for a long time and transported long distances in refrigerated tankers. The rate of use of natural gas has increased rapidly in the last decades. The world's production was 19 EJ in 1960, 46 EJ in 1975, and 58 EJ in 1982. The last value is about one-half of the oil consumption in the same year. (One EJ of heat is provided by 26.8 billion cubic meters of gas.)

The rate of use is expected to increase to about 80 EJ in 2000 and to peak at around 110 EJ in 2020. The most important sources of natural gas are summarized in Table 2.5.

The ultimate resources are known to a much lower accuracy, and the estimates have been rapidly moving upward as the interest in natural gas as a major energy source increases. Thus, for example, in three successive World Petroleum Congresses (1975, 1979, and 1983), the estimates of ultimate recoverable natural gas were given as 130 Tm³, 197 Tm³, and 271 Tm³, respectively. The last value corresponds to 168 GTOE (10^9 tonnes oil equivalent, or 7392 EJ) and is almost equal to the remaining reserves of conventional oil (see Table 2.1).

Large amounts of natural gas in various parts of the world are not amenable to exploitation by the usual techniques. Such gas is known as *unconventional natural gas*. The best-known, and partly exploited, is natural gas dissolved in interstitial water present in some formations under very high pressures (about 1000 bars) and elevated temperatures (100° to 200°C). These regions, known as geopressured zones, are found in many places in the world.

Another large potential source of natural gas is material locked up in porous sandstone formations that lack suitable fractures. Under such conditions the gas cannot migrate to traps where it can accumulate and from where it can be produced in the conventional way. To win this material, one must first fracture the gas-bearing formations. This is usually done by injecting very high pressure fluids into the rocks to crack them open and keep them open when the pressure is released. The gas seeps into the borehole through these cracks and is pumped away. Some experiments were carried out in the United States on the use of deep underground nuclear explosions to create large volumes of highly fractured gas-bearing rocks. The method showed some promise, but the experiments were discontinued because of the general opposition to nuclear explosions of any kind.

Another large potential source of natural gas is the gas hydrates. At temperatures close to freezing and under sufficiently high pressure (above about 25 bars), natural gas can form solid compounds with water. These compounds belong to the class of materials known as *clathrates*, which are three-dimensional assemblies of water molecules containing cavities of sufficient size to accommodate a methane molecule. The presence of the hydrocarbon molecule in the water cavities confers stability to the structure. In deep waters, where the temperature is usually a few degrees above zero, any methane formed in the sediments will combine to form a hydrate and thereby become trapped. Large amounts of gas hydrates have been detected as subbottom layers in many ocean sites. The total magnitude of this resource is still unknown, as are techniques for its exploitation. The magnitude of all unconventional gas resources is roughly estimated to be between 9000 and 31,000 EJ.

In addition to its obvious use as a convenient and clean fuel for electric power generation and industrial and domestic heat, natural gas is an important raw material in the chemical industry. It is the source of the important element hydrogen (from which ammonia for the fertilizer industry is made) as well as a whole range of chemicals.

Natural gas can be reacted with steam at high temperature and in the presence of suitable catalysts to give a mixture of hydrogen and carbon monoxide ("synthesis gas"). This important reaction is known as *reforming* and is the first step to many chemical processes, among which is the synthesis of methanol. Several commercial processes are available for methanol synthesis, and all involve catalytic reactions. The world's methanol production capacity in 1985 was estimated to be over 23 million tons.

Methanol is a low-boiling liquid and the smallest-molecular-weight alcohol; its molecule contains only one carbon atom. It is an important raw material in the chemical industry, but our interest in it at this point derives from its value as a possible alternative to gasoline (petrol). Methanol can be blended within certain limits with gasoline (about 5%) to stretch out this oil-based fuel. Certain complications in such blending arise from the mixture's incompatibility with water. Water will cause the blend to separate into two phases, so it is necessary to add cosolvents (higher alcohols or other oxygenated compounds) to the mixture. Methanol blends do have emissions and economical advantages and can be used in present automobile engines with no modifications.

Of greater interest as far as substitution of oil goes is the possibility of using almost pure methanol as a motor fuel ("neat methanol"). This fuel requires only minor modification of internal combustion engines and can be distributed through the existing system. A fleet of motor vehicles using neat methanol is now being tested in California, and results with respect to economic aspects and emissions are encouraging.

It is also possible to convert methanol to gasoline. A catalytic process has been developed by the Mobil Oil Co. using zeolites. This process converts methanol to high-octane motor fuel. Because methanol can be made from synthesis gas, which in turn can be made from natural gas or coal, this process can provide regular motor fuel from nonpetroleum sources. The first commercial plant to convert natural gas to methanol to gasoline using the Mobil process is operating in New Zealand. Its capacity is 1700 tons/day of gasoline, and it uses annually 1.5 million metric tons of methanol made from natural gas.

But for all its advantages, natural gas cannot be considered as a true long-range alternative to oil, simply because its known reserves and conjectured resources are not large enough.

2.5 Coal

Coal was the first fossil fuel used by mankind and will probably also be the last. Records of the use of coal as fuel go back to several centuries B.C. Coal is the result of biological, chemical, and physical processes that operate on plant material buried under different thicknesses of sediments. For the organic matter to escape oxidation by atmospheric oxygen, the initial conditions must be those of swamps with shallow water protecting the decaying plant material. Contemporary coastal swamps are probably a good approximation to the early stages of coal formation. In the past there were long epochs when in many parts of the world the climate was suitable for rich plant growth, and conditions existed for the preservation of the organic matter and its subsequent burial under thick layers of sediments. Although not limited to a particular geological period, the peak coal-forming time seems to have been 250 million years ago (the Carboniferous period). The prolonged mild heating of the sediments under pressure led to a gradual loss of the elements of water from the plant material and increased the proportion of carbon in it. Thus, peat has an elementary composition not much different from wood (cellulose): about 60% carbon, 34% oxygen, and 6% hydrogen. The lignites, which can be regarded as the next step in the evolution of coal, have 66% carbon, 28% oxygen, and 6% hydrogen. The bituminous coals are richer in carbon, analyzing at 80% to 90%, and the anthracites contain over 90% carbon.

Coal is not a single or simple chemical substance. The organic components of coal are mixed with various proportions of inorganic substances, which form the ash after combustion and are responsible for some of the sulfur emissions. The organic matter itself is a mixture of complex large molecules that varies according to the type and origin of the coal. These molecules may also contain part of the sulfur and nitrogen present in the original plant material.

Coals are ranked according to their calorific value. The major classes, starting from the highest grade, are the anthracites, whose calorific

Table 2.6. *Major world coal resources and reserves (in exajoules)*

	Proved amount in place[a]	Proved recoverable reserves[a]
United States	14,030	8,196
USSR	8,272	6,802
China	24,400	5,918
South Africa	2,543	2,120
Australia	2,543	1,696
Federal Republic of Germany	2,571	1,490
Poland	2,521	1,282
India	984	385
Canada	265	197
World	64,960	30,400

[a]Sum of the heat values of bituminous and subbituminous coals and lignites.
Source: World Energy Conference (1986).

value is about 33.7 GJ/tonne, the bituminous coals (36.3 GJ/tonne), subbituminous coals (28.5 GJ/tonne), and the lignites (or brown coals) with a calorific value less than 17.7 GJ/tonne. Coals from different localities, and from different seams in the same locality, may vary greatly in their properties. In addition to the differences in their heating value, coals differ in many other characteristics that may be very important for particular applications. Thus, coals may differ in the type and size of the flame they produce on combustion, the amount of gas produced on heating, their caking property, their sulfur contents, their ash content, and so on. Hence, not all deposits of coal are equally suitable for all applications.

Although coal is fairly widespread in nature, its distribution is far from uniform. About 88% of all the coal resources are concentrated in three countries (USSR, United States, and China), and a further 10% are distributed among six more countries (Australia, Federal Republic of Germany, United Kingdom, Poland, Canada, and Botswana). The remaining 2% is distributed among 23 countries. See Table 2.6.

The total reserves and resources (see definition in Section 1.2) shown in Table 2.6 are very large indeed compared with the combined values for oil and gas in Tables 2.1 to 2.4.

Less than a hundred years ago, coal was the major source of heat and power for industrial and domestic use. Coal was used directly to raise steam in industry, to generate electricity, to drive ships and trains, for cooking, and for domestic heating. Indirectly, in the form of coal gas made by the retorting of coal, it was used domestically for heating and cooking, for street lighting, and as a source of heat and power in industry. In most of these applications coal has now been displaced by oil or natural gas. The main energy use today of coal is in the generation of electricity.

The main nonenergy use of coal is as a reducing agent in the production of iron and steel. For this purpose the coal is first converted to coke by heating it to a high temperature in the absence of air. Only the better-quality coals are suitable for this application. The fraction of coal used in metallurgy was about 27% of the total (1977), but is now decreasing as more and more coal is being burnt to generate electricity. Coal is also being displaced in steel production by electricity and other reducing agents derived from oil or natural gas.

The fraction of coal use has dropped precipitously in the two decades preceding 1973 and is beginning to rise again. For example, coal accounted for 36% of the primary energy used in 1960 but only 25% of that used in 1978, illustrating its extensive replacement by oil and gas. The total global demand for coal, however, went on increasing slowly throughout this period. It amounted to about 54 EJ annually in 1960, 76 EJ in 1975, and 91 EJ in 1982. We have collected in Table 2.7 the change in coal use in the economy that occurred in OECD countries between 1960 and 1976.

The sharp decrease in coal use between 1960 and 1976 in the transportation, residential, commercial, and part of the industrial sectors should be noted. The use of coal in electricity generation, however, almost doubled in this period. These trends will probably continue, because they represent technological realities. Forecasts have been made of the use of coal in the energy sectors (excluding metallurgical and chemical uses) for the next few decades. Remembering the usual caveats regarding predictions and the scenarios on which they are based, we quote a few of them. For the year 2000 the estimates of coal use are in the range 125–150 EJ per annum, and for the year 2020 they range from 200 to 250 EJ (Frisch 1983).

Both the rate of use up to now and the forecasts for the next decades appear puny in comparison with reserves and resources listed in the

Table 2.7. *Consumption of coal by various sectors in OECD
countries (in exajoules per year)*

	1960	1976
Gas manufacture	0.845	0.202
Energy sector: use and losses	2.024	1.927
Total industry	9.755	8.017
Iron and steel	4.704	5.047
Others	5.051	2.970
Transportation	1.175	0.044
Residential/commercial	5.689	2.165
Electricity generation	10.023	18.709
Total coal[a]	29.718	31.143

[a]Totals include statistical differences.
Source: International Energy Agency (1978).

previous section. It would thus seem that the supply of coal is ensured
for many hundreds of years, and that all that is necessary is to switch
back to this primary fuel and thereby solve the energy problem for a few
centuries to come.

There are several reasons why this simplistic view may not be justified. For it is not at all certain that the known reserves and resources can
indeed be exploited fully, not only because of the many technological
problems that have to be solved before even a partial substitution of oil
and gas by coal becomes possible (these will be discussed fully in Chapter 6), but mainly because of the extensive damage to the environment
that very large scale exploitation of coal will entail.

As is seen in Table 2.6 the distribution of coal resources is very uneven, and it cannot be assumed that the few countries possessing most of
these will automatically undertake the role of the world's suppliers of
this commodity, particularly if the local environmental damage is perceived to benefit mainly a distant and foreign user. So that although
globally it may seem that the supplies are adequate, when regional or
national problems are considered the situation is not as clear.

A more severe environmental constraint to the continued large-scale
use of fossil-energy sources in general arises from the unavoidable emission to the atmosphere of carbon dioxide as the main product of the

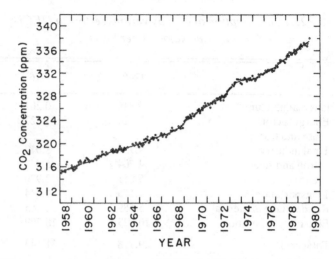

Figure 2.1. The increase in atmospheric carbon dioxide. (Source: Keeling, Bacastow, and Wharf 1982.)

combustion process. This substance absorbs infrared radiation, and its presence in the atmosphere reduces the escape of heat from the surface of the earth to space, thus leading to an overall warming of the globe, known as the *greenhouse effect.* In the absence of large-scale interference by human activity, the concentration of carbon dioxide in the atmosphere will maintain a steady state as a result of a balance between the rates of formation of carbon dioxide by combustion and respiration and of its incorporation into organic matter by photosynthesis (the carbon cycle). The injection of large amounts of carbon dioxide from the combustion of fossilized organic matter will lead to higher levels of carbon dioxide in the atmosphere. There is no doubt that this is already happening and that the concentration now (335 ppm) is higher than it was before the industrial revolution (see Figure 2.1). If the present trend continues, the concentration of carbon dioxide in the atmosphere is expected to double its present value by the middle of the next century (Trabalka 1985). There is little doubt that a general warming of the globe will occur as a result of an increase in atmospheric carbon dioxide, but the exact climatic effects of such a warming are far less certain.

The world's climate and global circulation are such complex subjects that a proper understanding and the ability to model them precisely, and thus predict accurately the effect of various factors, still elude us. At the present time the estimates of the average global warming range between 1.5° and 4°C. The warming will, however, not be uniform, and it is expected to be higher at high latitudes than at the equator. The temperature distribution and the pattern of rainfall and winds are sensitive to the details of the atmospheric circulation, and this is particularly difficult to model. Among the effects that are expected, without precise localization in space and time, are changes in rainfall. Some arid areas may then become fertile, and well-watered ones may become deserts. Another illustration of the possible catastrophic effects of the greenhouse effect is the chance that the Antarctic ice cap will melt during the warming, causing a rise of 6–8 m in the ocean's level. This will inundate many heavily populated areas around the world and appreciably change the shape of the coastlines.

In the absence of clear and unambiguous evidence of the deleterious effects of fossil-fuel combustion on global climate, it is difficult to formulate a clear policy in this regard. Even if clear evidence were available, it is doubtful that the political organization of the world would allow any such policy to be adopted, let alone be enforced. However, prudence dictates that until the global climate effects are properly understood, we must avoid excessive reliance on fossil fuels and devote maximum efforts to increasing the use of renewable energy resources and other zero-emission systems. These issues are discussed further in Chapter 7.

Additional reading

Grainger, L. and Gibson, J. (1981). *Coal Utilisation: Technology, Economics and Policy*. Graham & Trotman.

OECD/IEA (1982). *Natural Gas, Prospects to 2000*. OECD, Paris.

(1986). *Coal Information 1986*. OECD, Paris.

Trabalka, J. R. and Reichle, D. E., eds. (1986). *The Changing Carbon Cycle, A Global Analysis*. Springer-Verlag, New York.

Wilson, Carroll L., ed. (1980). *Coal – Bridge to the Future. Report of the World Coal Study*. Bellinger, Cambridge, Mass.

3

Nuclear energy

The extraction of useful energy from nuclear sources is a relatively new technology, and its practice is accompanied by many features absent in the case of fossil sources. Many environmental, political, and safety issues are involved, some of them subjects of great controversy. A proper understanding of the role of nuclear power requires a somewhat more detailed treatment of the subject than was sufficient in connection with conventional fossil technology. This chapter then reviews the main features of nuclear power generation.

3.1 Nuclear structure and isotopes

Atomic nuclei are composed of protons and neutrons held together by forces called, appropriately, *strong nuclear forces*. Since nuclei are electrically charged and often have magnetic moments, electromagnetic forces also play an important part in the energetics of the system.

Stable arrangements of the protons and neutrons (nucleons, in general) in the nucleus are only possible within very narrow limits of composition. Deviations from their optimal proportions give rise to nuclei that are unstable and undergo spontaneous changes to other more stable configurations. These changes may involve a conversion of one nucleon to another with the emission of particles and radiation. These unstable nuclei are then said to be *radioactive*. Radioactive nuclei greatly outnumber stable ones (2100 to 273).

The chemical properties of an element are determined by the magnitude of the electric charge of its nucleus (i.e., by the number of protons present in it). Over a narrow range, it is possible to vary the proportions of neutrons for a given number of protons. The several nuclei that result will have the same chemical properties (since they all will have the same number of protons) but will differ in weight and in most nuclear properties. They are known as *isotopes*. In some elements there may be several stable isotopes, but in all cases there will be unstable (i.e., radioactive) isotopes. For example, the element hydrogen consists of two stable isotopes (of mass 1, the nucleus of which consists of a single proton, and of mass 2, the nucleus of which contains one proton and one neutron). There is also an unstable (radioactive) isotope of hydrogen of mass 3 (the nucleus of which is composed of one proton and two neutrons). Because of the importance of these isotopes, they are given specific names: Hydrogen of mass 2 is called *deuterium* (D); hydrogen of mass 3 is called *tritium* (T). But, in general, isotopes are identified by their mass; thus ^2H is equivalent to D, and ^3H to T. Other examples of this nomenclature are ^{14}C for the radioactive isotope of carbon and ^{235}U for the light fissile isotope of uranium.

Because isotopes of the same element have very nearly identical chemical properties, their separation from the natural mixtures in which they are found is a very difficult process and is the basis of a major industry associated with the production of nuclear power. We shall say more about this subject later.

3.2 Radioactivity

Radioactive decay may involve the emission of electrons (known also as β particles), nuclei of helium (known also as α particles), and, more rarely, of neutrons and positive electrons. The electromagnetic radiation emitted is usually in the form of gamma rays and x-rays. Energetic particles and radiations are potentially damaging to biological systems because of the ionization and destruction of molecules that they cause. Hence, working with radioactive materials is hazardous.

The particles and radiations that are emitted in the course of radioactive decay carry off some of the energy of the parent nucleus. This energy appears as the kinetic energy of the emitted particle and the recoiling nucleus, or as the electromagnetic energy of the radiation. Ultimately the particles are stopped by the atoms around them, and the radiation is absorbed by the surrounding material, so the energy of the decay appears in the end as heat. An exception to this generalization is the energy carried off by neutrinos. Neutrinos are particles carrying no electric charge and possessing no (or very little) mass, which accompany β decays. They are not stopped by matter (on a terrestrial scale) and therefore escape with their energy. Thus a small fraction of the energy of radioactive decay and, as we shall see later, of fission escapes the confines of the earth.

A radioactive nucleus may change to a more stable one that is still radioactive, thus making possible a chain of radioactive decays until a stable nucleus is finally reached. The rate at which a radioactive material decays is a property of its nucleus and cannot be affected by human intervention. Because of the exponential nature of the decay of a radionuclide, the common measure of the rate of the process is in terms of its half-life. The *half-life* is the length of time needed for half the nuclei present at the beginning of the interval to decay. The half-lives of various isotopes cover a wide range of times, from a few milliseconds to billions of years. It is clear that only stable isotopes or those with very long half-lives can be found in nature from the time of the formation of the earth. Shorter-lived isotopes can only be found in nature if they are being continuously produced by some process. For example, uranium, thorium, and potassium-40, with half-lives of 4.5×10^9 years, 1.4×10^{10} years, and 1.3×10^9 years, respectively, are found in nature and are part of the primordial matter of the solar system. On the other hand, tritium (half-life 12.3 years) and ^{14}C (half-life 5730 years) are found in minute amounts in nature, because they are being formed continuously in the atmosphere by the action of cosmic rays. Many other radioisotopes of various half-lives are now being produced in large amounts as a byproduct of nuclear programs, both military and civilian. Some of these are the materials involved in the nuclear waste controversy.

3.3 Fission

The heavier elements are subject to another form of instability in addition to radioactive decay. The many protons in the nuclei of these elements favor their breaking up into two smaller nuclei. In this process, known as *fission,* a considerable amount of energy is liberated. For some nuclides this process occurs spontaneously, but for most others some degree of excitation of the nucleus is required before fission can take place. Among the naturally occurring nuclides, spontaneous fission is found only in thorium and uranium. Among the artificially produced transuranium elements, this mode of decay is quite common. The rate of decay of uranium and thorium by spontaneous fission is very low. For example, there is only about 0.5 fission per minute per gram of natural uranium. The rate of fission goes up dramatically, however, with the degree of excitation of the nucleus. Thus, the rate of spontaneous fission of ^{236}U from its ground state (unexcited state) corresponds to a decay half-life of about 2×10^{16} years. The same nucleus, when formed in an excited state by neutron capture in ^{235}U, fissions immediately (within 10^{-14} s). In connection with nuclear energy we shall be interested only in those fission events caused by the capture of neutrons in the fissile isotopes of uranium and plutonium: ^{233}U, ^{235}U, and ^{239}Pu.

Fission is not the only event that can occur upon the capture of a neutron by a heavy nucleus. The capturing nucleus may lose its excitation energy by the emission of a gamma ray and then live out the short- or long-life characteristic of its composition. The probability of occurrence of a particular reaction path is expressed by the term *reaction cross section* and is measured in barns, of dimension 10^{-24} cm^2. We distinguish, therefore, between neutron capture cross section, which expresses the probability of producing an isotope of the target nucleus one unit of mass heavier, and fission cross section, which measures the probability of a breakup of the target following neutron capture.

The various cross sections depend strongly on the energy (velocity) of the neutron. Therefore in dealing with the nuclear reactions involved in reactors, one must be careful to specify the energy of the neutrons involved – whether they are slow (2200 m s^{-1}), also known as thermal, or fast (about 1.4×10^7 m s^{-1}, corresponding to about 1 MeV energy). In Table 3.1 we show some of the capture and fission cross sections for the important fissile isotopes.

The precise mode in which a fissioning nucleus splits into two parts varies slightly from event to event, and so the fission products from many nuclei are distributed in a certain way around the most probable modes. Figure 3.1 shows the distribution of the products from the slow neutron fission of ^{235}U. Among the fission products are numerous radioactive nuclei of various half-lives. As Figure 3.1 shows, the symmetrical breakup into two equal fragments is not the most probable mode.

In addition to the nuclear fragments, several neutrons are also emitted during fission. Table 3.1 shows the average numbers of neutrons emitted upon the fission of various nuclides.

The fission of one uranium nucleus releases 201 MeV of energy. The MeV is the common unit of energy used by nuclear physicists and is equivalent to 1.60×10^{-13} J. To bring out the enormous magnitude of

Table 3.1. *Parameters for the reaction of neutrons with some fissile nuclides (cross sections in barns)*

	^{235}U		^{239}Pu		^{233}U	
	Thermal[a]	Fast[b]	Thermal	Fast	Thermal	Fast
Fission cross section	580	1.44	742	1.78	531	2.20
Capture cross section	98	0.22	271	0.15	47	0.15
Neutrons emitted	2.423	2.52	2.880	2.98	2.487	2.59
η[c]	2.073	2.19	2.108	2.75	2.284	2.43

[a] The thermal data are for 2200 m s^{-1} neutrons.
[b] The fast data are for reactor spectrum neutrons.
[c] The number of neutrons emitted per neutron absorbed.

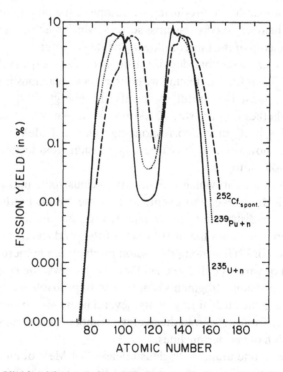

Figure 3.1. Yields of fission products from some fissile nuclides. (Source: Friedlander 1981.)

Table 3.2. *Distribution of energy released in the fission of uranium (%)*

Kinetic energy of fission fragments	83.5
Instantaneous gamma rays	2.5
Kinetic energy of prompt neutrons	2.5
Beta decay of fission products	3.5
Gamma rays from fission products	3.0
Neutrinos	5.0

this energy, we calculate that the fission of 1 g ^{235}U (containing 2.56 × 10^{21} atoms) will release 82.4 GJ or 22,900 kWh of heat (equivalent to the combustion of about 2 tonnes of oil). Most of the energy released on fission appears as the kinetic energy of the fragments (see Table 3.2). A fraction is emitted as electromagnetic radiation immediately on fission, and part is emitted upon the decay of the fission fragments. The kinetic energy of the fragments is converted rapidly to heat by their collision with atoms in the immediate vicinity. The electromagnetic radiation penetrates an appreciable thickness of material before being totally absorbed and degraded to heat. About 5% of the fission energy is emitted as neutrinos, which are not absorbed and escape into space, and their energy is thus lost to us. Table 3.2 shows the distribution of fission energy among the various modes.

3.4 Nuclear reactors

The circumstance that more than one neutron is emitted in the course of fission of a uranium nucleus opens the possibility of the controlled release of nuclear energy (and, of course, of nuclear explosions). If there were no neutrons produced in fission, or if the average numbers were less than 1, one would have to drive the fission reactions with external neutron sources. Although this is possible (for neutrons can be made readily in nuclear accelerators) the net energy balance of the system would be negative. That is, the energy that one would have to invest to produce the neutrons will be greater than the energy one would recover from the fissioning nuclei. This concept is being reexamined from time to time, as technology advances, as a method of producing fissile materials without the use of breeder reactors. The fact that, on the average, 2.5

neutrons are produced from each fissioning uranium nucleus (see Table 3.1) suggests that the process, once started, can proceed without the injection of outside neutrons. Moreover, it opens the possibility of generating an explosion of unprecedented power. This was realized very soon after the discovery of fission, and, the time being one of preparation for a war, the subject was pursued with great vigor. The results are too well known to need repeating here.

The conditions for using the neutrons produced in the fission itself to drive further nuclei to fission are fairly easy to deduce in a qualitative way. The precise calculations, which are complex, form the discipline of reactor physics and are outside the scope of this book.

To achieve a steady state of fission in a bulk of nuclear fuel, it is clearly necessary to ensure that each fissioning nucleus will cause precisely one other nucleus to fission (the system is then said to have a multiplication factor of unity). That means that out of the 2.5 neutrons produced, just 1 should be absorbed by another ^{235}U nucleus and cause fission. At first sight this seems simple, because one is starting off with such an excess of neutrons. But there are many ways in which the neutrons can be lost before they have a chance to cause fission. The conditions allowing a steady reaction to proceed are quite severe and constraining. The demonstration on December 2, 1942, that a steady nuclear fission chain reaction was possible is, therefore, considered a major step in the development of nuclear power.

We now come back to the conditions necessary for a steady-state reaction. There are two main causes for the loss of neutrons produced in fission: (1) The neutrons may simply escape from the apparatus before undergoing any reaction, or (2) they may be absorbed by the materials present (including the fuel assembly itself) without leading to fission. The first cause of neutron loss is minimized by providing a sufficiently large mass of fuel to reduce the escape probability. This minimal size, known as *critical mass*, depends on the composition of the fuel used and the configuration of the device. It may range over wide limits, from tons of natural uranium in graphite-moderated reactors to a few kilograms of ^{235}U in an atomic bomb. The second cause of loss is reduced by a suitable choice of materials, including those of the fuel elements themselves, and of their configuration. The need for enriched fuel in some types of reactors arises from such considerations.

In addition, means must be provided for the precise control of the number of neutrons present, for this determines the power level of the reactor. All these subjects form the art of reactor design. There are many ways of achieving a steadily operating nuclear reactor. Historically, each nation that embarked on nuclear power development started off from a technology it already possessed as a consequence of military nuclear programs. So in a sense the pattern of the civilian nuclear power programs was determined by the earlier military ones.

The United States emerged from World War II with the technology and facilities for producing fissile materials of high concentration. Plants existed for the production of ^{239}Pu and highly enriched ^{235}U. In addition, the development of nuclear propulsion for submarines led to the technology of light-water reactors (LWRs). It was natural therefore for the United States to follow a path of civilian power development involving enriched-uranium reactors and LWRs. All power reactors designed in the United States are based on enriched uranium. On the other hand, the United Kingdom and France emerged from the war with a general knowledge of production-reactor technology (i.e., reactors for producing weapon-grade plutonium) but with no isotope-enrichment facilities. It was natural therefore for both these nations to capitalize on their experience and proceed initially along the path of natural-uranium gas-cooled graphite reactors. Canada ended the war with great experience in the technologies of heavy-water production and natural-uranium processing and applied this advantage in the development of heavy-water natural-uranium reactors.

Power reactors of all these types were built and operated. The experience, both technological and economic, led gradually away from the natural-uranium–graphite design. Today, most of the reactors in the West are of the light-water–enriched-uranium type (basically the U.S. design). A surviving type is the Canadian CANDU, a heavy-water moderated reactor, which proved itself to be of a superior design. In the Soviet Union and, to a smaller extent, in Great Britain graphite-moderated reactors are still used.

Practically all power reactors in operation today (an exception being the prototype breeder reactors) are based on slow (or thermal) neutron fission. The probability of fission of a ^{235}U nucleus is a function of the energy of the neutron that hits it (this possibility is measured in terms of magnitude called the *cross section for fission;* see Table 3.1). It turns out that the probability of fission of ^{235}U is much greater for low-energy, slow neutrons than for fast ones. It follows that if we can cause the neutrons to slow down after being emitted in the fission process (or, as it is known, to "moderate" them), we shall greatly increase the chance of causing fission and reduce correspondingly their chance of escape.

The slowing down is achieved by allowing the neutron to collide several times with light nuclei before reaching a ^{235}U nucleus. Different reactor types use different light atoms as moderators. In LWRs the moderator is ordinary hydrogen in the form of ordinary water. In CANDU-type reactors, the light atoms are deuterium in the form of heavy water. In graphite reactors carbon atoms are used as moderators. Each type of moderator has its advantages and disadvantages. Thus, ordinary hydrogen is the most effective moderator, but has the disadvantage of absorbing some neutrons. This loss must be compensated for by an increase in the concentration of ^{235}U. Therefore, LWRs must use fuel containing 2% to 4% ^{235}U rather than natural uranium (which contains 0.7% ^{235}U). Although deuterium is somewhat less effective as a moderator, it absorbs neutrons very weakly, thus permitting the use of natural uranium as fuel. Its disadvantage lies in its cost. Heavy water must be made by a process of isotope separation and is therefore an expensive material. Graphite is a relatively poor moderator, requiring many collisions to slow down the neutrons. This in turn means larger volumes of material and bulky reactors.

From the foregoing it is clear that a nation aspiring for self-sufficiency in nuclear power must acquire and emplace an isotope-enrichment technology. It must either be able to enrich uranium and then employ LWRs, or it must produce heavy water and then be able to install reactors burning natural uranium. In general, the ability to enrich uranium means that weapon-grade material (highly concentrated ^{235}U) can also be made, hence the stringent controls exercised by the nuclear powers on this technology. We will say more about the safeguards issue later.

It was already mentioned that special provisions must be included in the reactor to ensure that the multiplication ratio of the neutrons is exactly 1. Usually this is done by control rods containing materials that absorb neutrons strongly. The reactor is then designed to have initially a multiplication ratio slightly greater than 1 with the control rods fully withdrawn, and slightly less than 1 with the controls fully inserted. It is clear that at some intermediate point the reactor will be just critical. The control rods are connected by a servomechanism to monitors of the neutron flux and maintain a steady level of the latter. Other safety features and multiplication factor controls are also used, but their discussion is outside the scope of this book.

We now briefly describe some general features of reactors. The main components, as far as the nuclear process is concerned, are the core, containing the fuel elements, the moderator (which may or may not also be the coolant), and the reactivity control mechanism. The main nonnuclear components are the heat removal system and the biological shielding.

We have seen that the net result of the nuclear fission process is the production of heat. So as far as power generation goes, a reactor is just a complicated heat source, analogous to the boiler of a conventional power station. Obviously, for the sake of increasing efficiency, one is interested in obtaining as high a temperature as possible out of this source. The upper limit is set by the materials of the reactor and its components. Most of the heat is generated within the fuel elements themselves, and their maximum permissible temperature is limited. The heat has to be removed efficiently in order to extract maximum energy within the permitted temperature range.

There are many configurations in which the components of a reactor may be assembled. We shall deal only briefly with some of them. In all cases the core has to provide the mechanical support for the fuel elements at the precise geometrical configuration required by the nuclear design of the reactor. At the same time the core has to allow for the free flow of the heat-removing fluid.

The coolant is frequently also the moderator, that is, it is ordinary water in LWRs. The two most common LWR configurations are pressurized-water reactors (PWRs) and boiling-water reactors (BWRs). Figure 3.2 is a schematic of these LWRs. As their names imply, the reactors differ in the way in which the heat is removed from the core. In the PWR heat is removed by liquid water under pressure, and no boiling is allowed within the reactor. The hot water under pressure generates steam in a separate device outside the reactor vessel. In BWRs the water is allowed to boil in the core, and the steam is led directly to the turbines. Most power reactors are one of these types, with the ratio of PWRs to BWRs being slightly greater than 2.

Only in graphite-moderated reactors is the coolant, of necessity, different. In the British and French designs the coolant is carbon dioxide gas, whereas the Russian design uses light water. Another type of gas-cooled reactor is the high-temperature gas-cooled reactor. It uses highly enriched uranium as the fuel and helium gas as the coolant. This reactor is designed to supply gas heated to over 900°C to be used to generate

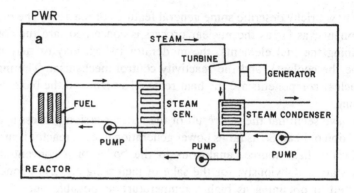

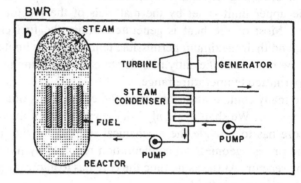

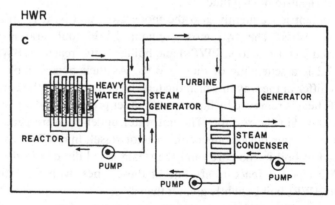

Figure 3.2. Various types of water-cooled and moderated reactors. (Based on
International Atomic Energy Agency 1982a.)

electricity more efficiently or as a source of industrial process heat. Only two such reactors are in commercial operation, one in the United States and one in Germany.

Recall from our discussion of the neutron economy in reactors that some of the neutrons are captured by the fuel without causing fission. These neutrons convert the uranium, after a short radioactive decay chain, to ^{239}Pu according to the scheme

$$^{238}\text{U} \xrightarrow{n} {}^{239}\text{U} \xrightarrow{\beta} {}^{239}\text{Np} \xrightarrow{\beta} {}^{239}\text{Pu}.$$

The fraction of the neutrons captured in this way to produce plutonium can be controlled to some extent by the design of the reactor and the materials used in it. This fraction is known as the *conversion rate* of the reactor. In LWRs about 0.4 atom of plutonium on the average is produced for each fissioning uranium nucleus. Some of this newly formed plutonium participates in the fission process and is burnt up, so in seasoned fuel elements the conversion rate is lower. In heavy-water reactors (HWRs), because of the much smaller neutron absorption of the moderator, the rate of conversion is higher and can reach 0.8. Reactors with very high conversion fraction are known as *advanced converters* and produce almost as much fissile material as they burn up.

It is possible to design reactors with a conversion ratio greater than unity – that is, reactors that make more fissile material than they burn in maintaining the chain reaction. Such reactors are known as *breeders*. It is quite easy to deduce the conditions for such a situation. All that is necessary is to ensure that each neutron absorbed by the fuel will produce more than two neutrons, one to maintain the chain reaction and one to make another fissile nucleus. Because some neutron losses are unavoidable, the required number of neutrons produced upon the absorption of a neutron by the fuel (known by the Greek letter η (eta)) must be appreciably greater than 2. Table 3.1 shows the basic parameters of the common fissile nuclei, including values of η. Examination of this table shows immediately that the most favorable system is plutonium based and operates with fast neutrons. One can also see that it is practically impossible to achieve breeding conditions with thermal neutrons and ^{235}U and ^{239}Pu systems. However, with ^{233}U, which can be derived from thorium by neutron capture and the radioactive decay sequence

$$^{232}\text{Th} \xrightarrow{n} {}^{233}\text{Th} \xrightarrow{\beta} {}^{233}\text{Pa} \xrightarrow{\beta} {}^{233}\text{U},$$

breeding may be just possible with thermal neutrons.

Breeder reactors essentially convert "fertile" material such as ^{238}U and

[232]Th into fissile material and thus make it possible to burn up practically all the uranium rather than the 1% or so that is possible in LWRs or HWRs. The installation of breeder reactors is therefore a very attractive proposition to countries that lack abundant uranium reserves. Moreover, because of the high utilization of the nuclear fuel, its cost presents a much smaller component of the cost of the power produced, and therefore the economics of the system is such that much more expensive uranium may be used. This feature opens up additional resources for nuclear power production. However, since the operation of a breeding reactor involves mandatory reprocessing and the handling of large quantities of plutonium (or [233]U), the issue becomes complicated by considerations of nuclear weapons proliferation and radioactive waste disposal.

3.5 Uranium supply

Uranium is fairly widely distributed in nature, and the average crustal concentration is about 2 ppm. Most of the primary uranium, however, is in the form of a fine dispersion of pitchblende (U_3O_8) or uraninite (UO_2) in igneous rocks. Rich veins of these minerals occur in several places in the world, but they are probably secondary in origin. The classical rich deposits in Czechoslovakia and Zaire are of this type.

Uranium has the chemical property that in its oxidized, hexavalent state it is rather soluble. When the atmosphere of the earth became oxidizing due to the presence of free oxygen (about 2200 million years ago), the leaching of uranium from rocks near the surface became possible. Surface and underground water streams carried small amounts of uranium, and when conditions were suitable the element precipitated to form secondary mineral concentrations. This process is still going on today and gives rise to tertiary and higher mineralizations. Most of the leached uranium finds its way to the oceans, giving rise to a concentration of about 3 ppb in seawater. The conditions for the precipitation of uranium in secondary forms are the presence of reducing agents (organic material, hydrogen sulfide) and of certain complexing materials, such as phosphates. Such conditions existed in many places over geological times, and this accounts for the widespread occurrence of uranium in sedimentary formations and its association with shales, coal, and phosphate deposits. In Table 3.3 we show the location of the most important uranium resources of the world. To the values shown in this table some

3.5   Uranium supply   45

Table 3.3. *Principal uranium sources (in thousands of tonnes,*
at less than U.S. $130 per kilogram)

	Reserves	Resources
United States	407	893
Australia	336	394
South Africa	313	147
Canada	185	510
Brazil	163	92
Niger	160	53
France	67	46
WOCA[a]	2,000	2,450

[a]WOCA = world, excluding centrally planned economies.
Sources: Nuclear Energy Agency/International Atomic Energy
Agency (1982, 1983) and International Atomic Energy Agency
(1980a).

1.5 million tonnes should be added to account for reserves and resources
in the centrally planned economies.

With uranium, an additional category of resources is in use, the *specu-*
lative resources. As the name implies, this category refers to uranium
thought to exist mostly on the basis of indirect evidence and general
geological considerations. It is estimated that this amounts to between
9.6 and 12.1 million tonnes in the world outside the centrally planned
economies, and to between 5.2 and 6.5 million tonnes in the centrally
planned economies. The total in all these categories has been estimated
at 27 million tonnes (Perry 1979).

All of these amounts refer to uranium that can be recovered at a cost
of less than $130 per kilogram (1983). But there are, in conventional
ores, large quantities of uranium that may become available at higher
cost. At least 2.3 million tons of uranium from conventional ores are
expected in the price range $130–$260 per kilogram. Although this is of
little interest at the present, future technological and economical developments may make them attractive.

In addition, much greater amounts of uranium can be obtained from
unconventional ores or recovered as by-products from other operations.

The uranium won from some of these unconventional sources is not necessarily more expensive than that obtained from conventional ores. We shall discuss some of the more important sources in the following paragraphs.

All marine phosphate deposits contain some uranium. The concentrations range from 10 ppm to as high as 700 ppm, but a good average figure is around 100 ppm. The world's reserves of phosphate rock are estimated at about 130,000 million tonnes, containing therefore about 13 million tonnes of uranium. The extraction of uranium from phosphate rock as a sole product is not yet economical (about $500 per kilogram). Uranium may, however, be obtained readily as a by-product of the production of phosphoric acid by the wet process. But only a small part of the phosphate rock mined today is so processed. If all the present production of phosphoric acid by the wet process were associated with uranium recovery, over 6000 tonnes/yr could be obtained. Plants for the separation of uranium as a by-product of phosphoric acid manufacture exist in the United States, Canada, France, and other countries. The total production of uranium from this source in 1980 was estimated at 1000 tonnes.

Uranium is also found in marine shales, with concentrations ranging from 10 to 80 ppm. An unusually uranium-rich shale deposit occurs in Sweden (300 ppm U). A well-known potential source of uranium is the Chattanooga shale formation in the United States with a concentration of 60 to 80 ppm and a total amount of some 10 million tonnes of uranium. Uranium is also found in some low-grade coals and lignites.

Another source of by-product uranium is the copper industry. The United States, Chile, and South Africa are possible sources for such uranium.

Although most igneous rocks have a low uranium concentration of 2 to 4 ppm, some granites have much higher values. For example, the Conway formation granites in New Hampshire contain 12–15 ppm of easily leachable uranium, giving a resource of over 5 million tonnes. Whether these low-grade, huge resources will ever be exploited is open to question. Even if technological developments make this high-cost uranium economically feasible, the severe environmental disturbance involved in processing vast amounts of rock may be unacceptable.

The oceans, with a very low uranium concentration of about 3 ppb, contain some 4000 million tonnes of uranium. Processes have been elaborated for the extraction of uranium from seawater, and a pilot plant (100 g/day) has been erected in Japan. Costs are high, between $500 and $1000 per kilogram, and are unacceptable for present technologies. This source of uranium may be of interest to future technologies involving breeders.

Although there seems to be no major environmental problem associated with the extraction of uranium from seawater, the technological difficulties are formidable. The huge amounts of water that have to be handled rule out any process involving pumping. Instead, use would have to be made of tides and ocean currents to deliver the water to the processing units. A major consideration in the choice of a suitable location for such a process is the avoidance of recycling water from which the uranium has already been extracted. Several locations exist that may be suitable for such uranium extraction.

3.6 Thorium

Although at present very little thorium is used to fuel power reactors, future technologies may be based on this element. Thorium can be used as the fertile material in breeder reactors, in high-temperature gas-cooled reactors, and in HWRs. The lack of demand so far has meant a correspondingly small effort at prospecting for this element. It is therefore likely that much more thorium is available.

The chemistry of thorium differs from that of uranium in that a relatively soluble hexavalent state is absent. As a consequence, the whole series of secondary and higher mineralizations caused by leaching and reprecipitation is missing in its geochemistry. Thorium is found therefore mainly dispersed in igneous formations and as a mechanically reconcentrated weathering product of such rocks. Thus, in many parts of the world thorium is found in the heavy components of black beach sand that has been sorted out by wave action and in placer-type deposits in ancient river beds where it has been concentrated by the action of water. Thorium is also found in vein formations and in pegmatites and conglomerates. In Table 3.4 we show some of the principal known sources of thorium.

3.7 Fuel cycle

Unlike other forms of primary energy, considerable processing and manipulation of the nuclear fuel is necessary before energy can be extracted from it. The fuel cycle begins with the mining and milling of the uranium

Table 3.4. *Major thorium deposits (in thousand tonnes)*

	Reserves	Resources
Turkey	330	440
India	319	NA
Norway	132	132
United States	122	278
Brazil	68	1200
Greenland	54	32
Egypt	15	280
Canada	NA	300
Others	33	90
Total	1,073	2,752

Source: Nuclear Energy Agency/International Atomic Energy Agency (1982, 1983) and International Atomic Energy Agency (1980a).

ore. It is then treated with acid to dissolve the element. For poor ores, a process known as *heap leaching* is sometimes used. In this process piles of the crushed ore at the mining site are sprinkled with dilute sulfuric acid, which is recirculated until the uranium concentration in the solution reaches a few percent. In all cases, the uranium solutions are purified, and the uranium is precipitated as "yellow cake" (crude uranates), which is the form in which uranium is usually shipped. For nuclear use the uranium must be very pure and particularly free from neutron-absorbing substances. In the fuel factory, therefore, the yellow cake is dissolved and purified, usually by a process involving solvent extraction. If the uranium is destined for the fabrication of natural-abundance reactor-fuel elements, the purified material is converted to the dioxide UO_2. If the uranium is destined for isotope enrichment, it is converted to the hexafluoride UF_6. Uranium hexafluoride is an unusual material, being a very volatile solid (boiling point at atmospheric pressure is 56.2°C), and is stored and shipped in gas cylinders. The close similarity in the properties of isotopic molecules makes the separation of the naturally occurring mixture into two fractions, one enriched in the desired isotope and the other depleted with respect to it, a formidable technical problem. The effort involved in the separation of isotopes is expressed in special units called *separate work units* (SWUs). Their value is expressed as kilograms separation work and depends on the concentration of the product and the concentration of the "waste," or depleted, uranium. For example, the production of 1 tonne of 3% ^{235}U from natural uranium (containing 0.714% of ^{235}U), leaving depleted material of 0.2% ^{235}U content, requires 4290 units of separative work and 5.5 tonnes of natural uranium. The production of 1 kg of 93% ^{235}U from the same starting material and leaving the same waste concentration requires 235 units of separative work (which may be compared with the 4.3 SWU/kg needed for the 3% material).

Although HWRs can use natural uranium, LWRs require a ^{235}U concentration between 2% and 4% (three- to fivefold enrichment). For weapons use, the concentration required is above 93% ^{235}U.

Two principal methods are used today to enrich uranium on the commercial scale: gaseous diffusion and gas centrifugation. In the gaseous diffusion process, use is made of the fact that the molecule $^{235}UF_6$ is about 1% lighter than $^{238}UF_6$ and therefore its rate of diffusion through a porous membrane is slightly higher (1.0043 times that of $^{238}UF_6$). By repeating the process many times, we may reach any desired concentration. The process was developed in the United States during World War II for the production of weapon-grade ^{235}U. The plants are huge and consume staggering amounts of power, mostly in recompressing the UF_6 gas between diffusion stages. Diffusion plants exist in the United States, the United Kingdom, France, the Soviet Union, and China.

Gas centrifugation, in its initial development, can be traced even to prewar years, but it did not become commercial until the 1970s. In this process use is made again of the difference in mass between $^{235}UF_6$ and $^{238}UF_6$ molecules. The mixture of gases is spun at very high rates of rotation in cylindrical gas centrifuges. The heavier molecules tend to concentrate near the outside wall, leading to a small enrichment in the radial direction. An internal circulation of gas is induced in the centrifuge, which leads to the multiplication of the effect in the axial direction so that a much greater effect is obtained between the two ends of the device. Even then it is necessary to repeat the process many times in a cascade of centrifuges to obtain the desired enrichment. Because this process is far more economical in power consumption, it has become the method of choice for new construction. Gas centrifuge plants exist in the United States, United Kingdom, and Holland (a consortium of the United Kingdom, Holland, and Germany). It is reported that a gas centrifuge plant also exists in Pakistan.

Another class of processes that received some attention is based on slight differences in the aerodynamic behavior of isotopic molecules. In one process, developed in Germany (Becker), uranium hexafluoride mixed with hydrogen is allowed to expand through a specially shaped nozzle to yield two streams of gas differing somewhat in isotopic composition. The process has to be repeated many times to reach the desired concentration. A prototype plant was constructed in Brazil.

In another process of this type developed in South Africa, uranium hexafluoride mixed with hydrogen is allowed to expand in a vortex device, again leading to enriched and depleted streams. A prototype plant is in operation in Palindaba.

All these processes may be superseded by a new method being developed in the United States, France, Germany, and Japan. According to this entirely new approach, laser beams are used to selectively excite and ionize ^{235}U atoms in uranium vapor. The ions, highly enriched in ^{235}U, are collected by electromagnetic fields. The advantage of this method lies in the high separation obtained in each stage. Very few stages are therefore sufficient to reach high concentrations of ^{235}U. The process also appears to consume less power than others. The development seems to have reached a sufficiently convincing state for the United States to decide in 1985 to stop halfway the construction of a new, large, gas centrifuge plant in Portsmouth, Ohio, and to rely for future expansion on the new laser technology.

3.7.1 Fuel elements

Whether the nuclear fuel is natural or enriched uranium and whether its chemical form is a metal or an oxide, it must be fabricated in a form suitable for use in a reactor. The demands on such a fuel element are quite severe:

1. It must be capable of efficiently transferring the large amount of heat generated within the fuel to the coolant outside.
2. It must protect the nuclear fuel from attack by the coolant (most frequently water at high temperature and pressure).
3. It must contain absolutely all the fission products generated within it.
4. It must withstand the severe radiation flux over long periods without damage.

All these requirements are difficult to achieve, and the success of the nuclear industry is in no small measure the result of the satisfactory solutions of these problems.

Satisfactory fuel elements have been produced by paying attention to the material properties of the fuel itself – for example, by developing special alloys for the metal or ceramic forms for the oxide fuels. The other important component of a fuel element is its casing or cladding. Its function is to protect the fuel and to prevent any radioactivity from escaping into the system. The casing usually consists of a close-fitting sheath of stainless steel or an alloy of zirconium (Zircalloy) hermetically sealed around the nuclear fuel.

The maximum length of time a fuel element can be exposed in a reactor depends on its physical integrity and on changes in its nuclear properties. The first factor is rather obvious. Even the best-designed fuel elements suffer some radiation damage and fatigue, which accumulate with time to the point where their integrity is endangered. Rather than risk failure and the consequent radioactive contamination, technicians replace the element. The other limiting factor arises from the fact that some of the fissile material in the fuel is burnt up and that an equivalent amount of fission products appears. Some of these fission products have very high neutron absorption cross sections and compete effectively with the fissile nuclides for the neutrons. The net result of these changes is that the reactivity of the reactor goes down and a point is reached where this decrease cannot be compensated by the control rods. Part, or all, of the fuel loading must then be changed. In PWR-type reactors, for example, one-third of the core is replaced by fresh fuel every year.

It is clear that there is great economic advantage to extending the maximum burnup limits as high as possible, for the load on the whole fuel cycle decreases in proportion. Modern LWRs are designed for a burnup of about 30,000 megawatt-days (MWD) per tonne. Breeders and high-temperature gas-cooled reactors have achieved burnups of over 100,000 MWD/tonne. To illustrate the magnitude and implications of such burnups, recall that in each tonne of fuel 100 kg (10%) are fissioned, that 10% of the atoms of the fuel have been replaced by twice the number of a mixture of completely unrelated atoms, and that all this is supposed to happen without affecting the integrity of the fuel element.

3.7.2 Storage and reprocessing

Used fuel elements are intensely radioactive and generate a good deal of heat as the result of the decay of the fission products present. They must therefore be cooled even when removed from the reactor and kept in storage, and the surroundings must be protected from the intense radiation (IAEA 1980e). These aims are usually achieved by storing the spent fuel elements under water in pools provided in the reactor building. After a sufficiently long time the cooling may be dispensed with, but the radiation protection is required "forever."

The spent fuel elements contain valuable materials: the unburned part of the original fissile element content and usually a considerable amount of additional fissile material produced as the result of neutron capture by the fertile materials (uranium or thorium) in the fuel. The uranium content of the fuel may also have some value. Whether to reprocess the spent fuel and recover the valuable components or to leave the fuel in storage forever is a complex and controversial problem: In addition to its economical aspects, weapons proliferation and radioactive waste disposal must be considered. We shall return to these issues later, because they have a great impact on the future development of nuclear power (IAEA 1980c).

After cooling for one to three years the spent fuel elements may be sent for reprocessing if so desired. In the reprocessing plant the fuel is chopped up and then treated with nitric acid to dissolve the uranium oxide. During this operation the gaseous fission products are released. These contain radioactivities due to ^{129}I and ^{85}Kr, which must be trapped to prevent their escape to the atmosphere.

From this point on the process may continue along several possible paths. The most commonly used procedure for treating LWR fuels is the Purex process, which will be outlined later. This process separates the plutonium, uranium, and fission products present in the solution of the fuel elements into individual streams. The chemistry involved is not too complicated, and the procedures are based mostly on solvent extraction. The unit operations involve pulsed columns, mixer–settler devices, and centrifugal contactors. If it were not for the intense radioactivity of the materials handled, the process would be regarded as fairly straightforward. However, the high radiation fields and the toxicity of the materials dictate operation by remote control from behind thick biological shields. Because of the difficulties of carrying out routine maintenance operations under these conditions, the plant must be designed in such a way as to minimize the need for direct access. Moving parts in the critical locations are avoided, and as many components as possible are located outside the shield. Remote maintenance facilities are also provided in the high-activity areas. All these features make the reprocessing of spent fuel elements one of the most complex and sophisticated chemical engineering processes. It is also a most expensive one.

The nitric acid solution of the fuel element is subjected to solvent extraction, which results in a pure plutonium nitrate solution, pure uranium solution, and a highly radioactive solution containing the fission products. In addition, considerable volumes of medium- and low-level radioactive solutions are generated. The plutonium and uranium are then usually precipitated as the dioxides. The plutonium is either sent to storage or to fresh fuel fabrication. Reprocessing plants exist in the United States, United Kingdom, France, Belgium, Germany, Soviet Union, India, and China. Some of these countries provide custom reprocessing service.

The main motivation at present to reprocess fuel is to recover the plutonium content. This is either stored for future use in reactors, or, failing a world at peace, is taken for weapons production. In the United States, for example, all the reprocessing done so far was for military applications, and no commercial reprocessing facilities exist. Weapons-grade plutonium, in general, requires lower burnups, so specially designed reactors and fuel cycles are used for this purpose. Most of the plutonium separated from commercial power reactors is stored probably for future use in breeders.

3.7.3   Radioactive wastes

Recall that in the fission of uranium or plutonium, over 70 isotopes of some 26 elements are formed. Most of these are radioactive, and some have very long half-lives. So long as the fuel elements remain integral, the fission-product radioactivity is locked within them and cannot escape into the environment. However, as soon as the spent fuel elements are reprocessed, the radioactive products become mobile in the form of solutions and gases and their disposal becomes a major problem (IAEA 1980f).

In reprocessing, the fission products are separated out and become the high-level radioactive waste streams. In addition, there are streams of medium- and low-level activities. Even if the spent fuel is stored for one or two years prior to reprocessing, the level of radioactivity is still very high. With high-level waste, the activity is so high that cooling is necessary to remove the heat generated by the radioactive decay of the fission products.

In addition to the fission products, the waste streams contain small amounts of plutonium and transplutonium elements. Some of these are α emitters and extremely hazardous. So, altogether, the waste streams from reprocessing operations are unpleasant and dangerous materials, and the problems involved in their handling and disposal have challenged the nuclear community from the beginning. The problems are not yet entirely solved.

As a temporary measure and to allow for further decay, the solutions are usually stored in special tanks that are carefully monitored for leaks. The main radioactivity at this stage is due to isotopes of strontium and cesium (with half-lives of about 30 years) and the actinides (with half-lives extending to millions of years). Keeping the solution in tanks for thousands, or even hundreds, of years is an impractical proposition and therefore means must be found for the more permanent disposal of these materials. This problem has been the center of public debate for many years. It is aggravated by the fact that once the radioactive materials get into the environment they will ultimately distribute themselves throughout the globe and affect all life, not only in the country where they originated but also elsewhere. As a result, what happens to these materials in each country is of international concern. Various regulations and international standards have been imposed on the nuclear industry to minimize the danger from reprocessing wastes. How effective they are remains to be seen.

On the technical side, the problem is to immobilize the radioactive materials in a suitable solid form and thus prevent them from getting into the environment. The medium-level activities are usually solidified with concrete or asphalt and then stored in special locations. The low-level activities are either discharged into the sea (there is a good deal of unhappiness about this practice) or concentrated first and added to the medium-level activity.

It is the high-level wastes, however, that present the greatest challenge. The present approach is to immobilize them by fusing with a suitable matrix to form a glass (or artificial rock) and casting them into cylinders. One seeks compositions that will be stable over very long times and from which the activities cannot be leached out. The active cylinders are further encased in metal and are stored underground.

The choice of underground storage locations also presents a difficult task. Because one is seeking sites that have to be safe over periods longer than the history of mankind, one has to look for locations that have proved to be stable over geological times and that are far from groundwater. The favorite locations, where they are available, are deep salt deposits. In addition to many technical advantages and conveniences, the mere existence of a salt deposit is a guarantee of its isolation from groundwater and the environment. Other suitable burial locations are in caverns excavated in hard crystalline rocks. We re-emphasize that the problem of high-level wastes faces only those nations that include reprocessing as part of their fuel cycle. If a nation chooses a once-through cycle for its reactors, the problem does not exist. It is faced instead with the need of storing many spent fuel elements for a long time. Guaranteeing the integrity of spent fuel elements over geological times is also not a simple task.

The limitation to a once-through system, however, is a very severe one regarding the possible contribution of nuclear power to the energy needs of mankind. As will be shown later, it means a much shorter life for this energy source and makes it doubtful whether the effort is worthwhile. As a consequence, one must accept the fact that there will be strong pressure to reprocess the spent fuel in order to recover the fissile and fertile components and recycle them in breeders or other fuel-saving cycles. Therefore an economically and socially accepted solution to the "permanent" storage of fission products and other activities is a necessary condition for an extended contribution of nuclear energy to human needs.

3.8 Fusion

The discussion of fusion energy is included in this chapter because fission and fusion energies have the same origin. However, fusion power is as yet an unproven technology, and its discussion will therefore be more in the nature of a description of the principles involved, the problems to be solved, and the directions of present research efforts. It was pointed out in Section 1.1 that fusion power is derived from a source that is not really renewable, but its magnitude is such as to make it practically limitless.

Fission and fusion energies originate from the rearrangement of the nucleons from which the reacting nuclei are composed. In fission, a complex and relatively large unstable nucleus (uranium, thorium, or plutonium) breaks up into smaller nuclei with a release of energy. In fusion, two light nuclei join to give a heavier and more tightly bound nucleus with the release of energy. In terms of the energy released per gram of material, fusion sources are more prolific than fissile materials. To be more specific, in fusion reactions we deal with the combination of isotopes of hydrogen to give nuclei of helium. Several reactions are possible, and because they have somewhat differing raw materials we enumerate some of them:

$$^2\text{H} + {}^2\text{H} \rightarrow {}^3\text{He} + {}^1n + 3.3 \text{ MeV},$$

$$^2\text{H} + {}^2\text{H} \rightarrow {}^3\text{H} + {}^1\text{H} + 4.0 \text{ MeV}.$$

We shall refer to these reactions as the D-D fusion.

$$^2\text{H} + {}^3\text{H} \rightarrow {}^4\text{He} + {}^1n + 17.6 \text{ MeV}.$$

We shall refer to this reaction as the D-T fusion. The isotope of hydrogen, tritium, is not found in nature in any practical abundance. It can be made from lithium and neutrons by the reaction

$$n + {}^6\text{Li} \rightarrow {}^3\text{H} + {}^4\text{He}.$$

The isotope of lithium ^6Li is found in nature with an abundance of 7.5%, and the neutrons for the preceding reaction can be produced in the D-T reaction. Because these neutrons are very energetic (14.1 MeV) and some multiplications of their number by $(n, 2n)$ reactions will occur, matters can be arranged so that more tritium is produced than is consumed in the D-T reaction. In other words, tritium breeding is a possibility. The D-D and the D-T reactions will only take place if the nuclei are forced together to within a certain minimal distance. To do this the nuclei must possess sufficient kinetic energy to be able to overcome the mutual electrostatic repulsion and lead to the desired reaction. This energy is not very large (some tens of kilovolts) and is easily attained by accelerators. However, no practical fusion energy device can be based on particle accelerators because the net energy obtained will be a small fraction of the energy invested in the acceleration process.

An approach to the practical release of fusion energy, in an amount greater by a considerable margin than the energy invested in the process, is to heat the mixture of reactants to a very high temperature. At high temperatures some of the atoms or ions (because the gas will be totally ionized under such conditions) will possess sufficient kinetic energy to undergo the fusion reactions. The temperatures required for this to happen are measured in hundreds of millions of degrees. The temperature for ignition, that is, the temperature at which the rate of energy production by fusion is just balanced by that lost by radiation, is about 40 million degrees for the D-T reaction. At these high temperatures all matter is in an ionized gaseous state, known as plasma, and there is no known material vessel that can contain the reaction mixture. The main problem in fusion research is, therefore, to find means of handling and confining a plasma for a long enough time for it to generate more energy than was invested in producing it. The break-even point will depend on the product of the number of fusion reactions that take place per second (i.e., on the density of the plasma and its temperature) and the length of time the plasma is held together. This product is known as the Lawson criterion and is equal to 2×10^{14} cm^{-3} s for the D-T reaction at 10 keV and 5×10^{14} cm^{-3} s for the D-D reaction at 100 keV.

Many systems for achieving plasma confinement for short periods have been studied, and work on some of the more promising concepts is still continuing. One can divide the approaches into two. The more common one is to rely on magnetic fields of special configuration to prevent the plasma from dispersing too soon. There are at present at least 10 different major methods being investigated using this approach. In addition, there are several ways of achieving plasma heating, so the number of promising combinations is great, and this fact explains the many experiments under way in laboratories around the world. Another, and altogether different, approach is to dispense with magnetic confinement and instead rely on the inertia of the components to keep them close together long enough to generate a considerable amount of energy. This approach approximates closely to miniature hydrogen bombs. There are at least four different proposals for achieving this. The best known of these systems uses a powerful laser beam to generate a micro thermonuclear explosion in a tiny pellet of D-T fuel. Useful power would be generated by carrying out a continuous sequence of such microexplosions in a suitable reaction chamber where tritium would be bred and from which the energy would be removed as heat to drive turbines.

Major fusion projects, mostly based on the magnetic confinement approach, are under way in the United States, Soviet Union, United Kingdom, France, Germany, Japan, and other countries and involve large budgets. It is turning out to be a most frustrating field of research with new and unexpected phenomena cropping up whenever the frontier of knowledge is pushed ahead a bit. For this reason the time schedule is continuously sliding forward, and the initial optimism has given way to much greater caution. The proof of feasibility still eludes the many teams seeking it, and it seems that there is still a long way to go before commercial application will become possible.

Once the feasibility has been demonstrated, the need will arise for very extensive engineering development, because the conditions for a practical fusion reactor are very different from those familiar today in the nuclear industry. New materials will have to be developed that will perform well under the severe conditions expected to be found in fusion reactors, and new procedures and techniques will have to be perfected. It is therefore not very likely that fusion power will contribute appreciably to the energy substitution process until well into the next century, if not longer. If and when it does, its contribution will be to the electrical sector, and it will be in competition with fission energy. Many of the problems that now bedevil the fission-based nuclear energy industry will also trouble the fusion energy. The large minimal size of the fusion reactor will limit its field of application and will involve a high level of investment. Fusion technology will not be free from the problems of radioactive wastes, although they will be different in kind and less severe than those of fission. Instead of fission products there will be a variety of activation products from the various materials of construction, some of fairly long half-life. Instead of plutonium there will be a large amount of tritium in the fuel cycle. This certainly has a smaller weapons-proliferating potential, but it is not negligible. As far as the danger of the misuse of the knowledge gained through the application of controlled fusion to weapons development is concerned, the inertial confinement technology is probably a sensitive area. The initial hope that in fusion there will be found a "clean" energy source free from environmental and safety problems has not yet been realized.

So far we have not distinguished between the two kinds of fusion reactions that we described earlier. There are, however, profound differences. The most important one is that they differ greatly in the ease with which they proceed and the minimal temperatures needed to achieve ignition and break-even conditions, about 300 million degrees for the D-T reaction and about 1.7 billion (10^9) degrees for the D-D reaction. The D-T reaction is, therefore, by far the easier to carry out, and for this reason practically all the work carried out in the world on fusion and most of the data that are presented relate to this reaction.

There is also a great difference between the two processes in their dependence on raw materials. The D-D reaction is based entirely on deuterium, universally present in all water sources (1 part in 6700 of all hydrogen in nature is deuterium), and processes for its extraction on a large scale are now proven technology. The amount of energy released upon the complete fusion of 1 g of deuterium is 96,000 kWh(th). The total deuterium content of the oceans (23 million tonnes) can therefore yield 2.2×10^{24} kWh(th) or about 8×10^{12} EJ. This prodigious amount of energy is about 26 billion (10^9) times the present annual world consumption. It is sufficient for much longer than the solar system is expected to last. The cost of deuterium is also relatively low, and if its burning became practical, the fuel component of the electrical energy cost would be negligible. For these reasons D-D fusion is an attractive long-range solution.

The D-T reaction depends for its raw material on lithium from which the tritium is bred in the fusion reactor. Lithium is fairly widespread in nature, although concentrated minerals of this element are rarer. The amount of lithium needed to produce a thermal megawatt in a D-T fusion reactor depends heavily on the detailed design; it could vary between about 10,000 kWh(th)/g lithium down to about 6000 kWh(th)/g. The world's resources of lithium are not known precisely mainly because of the limited interest in this element. However, it is estimated that the terrestrial resource amounts to some 84 million tonnes. In addition, lithium is present in ocean waters at a concentration of 0.17 g/m^3, and, if a 50% extraction efficiency is assumed, the magnitude of this resource is some 120 billion (10^9) tonnes. If we combine these amounts with the lower of the thermal fusion yields, we obtain an energy potential of 7.2×10^{20} kWh(th) or 2.6×10^9 EJ. This amount is approximately 9 million times the present total annual world energy consumption. Also in this case the cost of the lithium would contribute an almost negligible fraction to that of the electricity produced.

One therefore sees that even without the more difficult deuterium-burning, fusion has an energy supply potential well beyond any planning horizon, greater than the potential of uranium with breeder technology, estimated at 190 million EJ (some 600,000 times present world annual energy consumption). It is obvious that with numbers of such magnitude the difference between the potential of the various sources is irrelevant to any conceivable policy discussion. These numbers explain the motivation behind the extensive research and development effort on fusion in various parts of the world. Current programs are running at the rate of probably $1.5 billion (American) a year worldwide. This is a remarkable feature for such a long-range project and shows the importance attached to a possible permanent solution to the energy needs of mankind.

Even if fusion is technologically and economically successful, there is another prerequisite it must meet before it can meet all, or most, of mankind's needs – the ability to substitute for other fuels also in the nonelectric sector. We shall discuss this question later.

Additional reading

Evans, N. L. and Hope, C. W. (1984). *Nuclear Power: Futures, Costs and Benefits*. Cambridge University Press, Cambridge.
Rippon, S. (1984). *Nuclear Energy*. Heinemann, London.

4

The role of nuclear power

We have not found it necessary to discuss the role of each of the fossil fuels as a separate issue, because in their case the main forces that determine their share of the energy supply are largely economic. In the nuclear case, however, there are many factors, in addition to obvious economic ones, that have a major influence on the role this source of energy can play. These factors extend from the technical characters peculiar to nuclear energy, to environmental effects, and to social and political considerations. It is for this reason that a special chapter is devoted to a discussion of the possible role of nuclear energy.

4.1 Nuclear fraction

At present, nuclear energy is used almost exclusively for the generation of electricity, and this fact immediately limits our discussion of its role in the energy economy to that sector. Its share of the total energy market is therefore closely tied to that fraction of the energy needs of the economy that is supplied by electricity. In Table 4.1 we show this fraction for various regions of the world.

We note from this table that in the industrialized countries about a third of the primary energy is used in generating electricity, whereas for most developing nations the proportion is less than a fifth. The table also shows the proportion supplied by nuclear energy, which is quite small.

We may now ask, what fraction of the total electrical energy is it reasonable to expect nuclear energy to provide? To answer this question, we must examine briefly the way electrical networks are designed and managed. In any country where a choice exists, there is an optimum

Table 4.1. *Percentage of primary energy used to generate electricity and the nuclear contribution, 1985*

	Total energy consumption (EJ)	Percent used for electricity generation	Percent nuclear
North America	81.9	34.2	5.2
Western Europe	51.9	35.0	10.2
Industrialized Pacific	19.7	39.4	7.4
Eastern Europe	77.6	25.8	2.5
Asia	49.2	17.3	0.9
Latin America	19.5	22.1	0.4
Africa and Middle East	19.1	19.4	0.3
World total	318.8	28.4	4.2
Industrialized countries	214.7	32.7	5.9
Developing countries in CPE Europe	14.7	23.3	2.0
Other	89.4	18.7	0.6
Total developing countries	104.1	19.3	0.8

Source: International Atomic Energy Agency (1986).

mix of the primary energy sources for electricity production that yields the most economic and reliable electricity distribution system.

Each primary energy source has its own characteristics and its particular advantages and disadvantages. The determination of the optimum mix is often complex. An important factor is the daily and seasonal pattern of the load on the system. This is usually presented either as a load curve showing the system's load as a function of time, or as a duration curve showing the fraction of time during which a certain load is exceeded. The fraction of the load that is always present (i.e., always exceeded) is known as the *base load*. This fraction varies from system to system. Demand near the maximum, or peak, is present only for a small fraction of the time – a few hours a day or a few hundreds of hours a year.

The hierarchy of the use of thermal electricity generating units in any network will, in general, involve using the large, modern stations to supply the base load. The older, smaller, less efficient, and somewhat more flexible units will be used as spinning reserve and suppliers of partial demand. Special, very flexible, and relatively low cost (small) units burning expensive fuel will be used to satisfy the peak load. Gas turbines are usually used for this service. Hydropower, if available, is also sometimes used for peak supply.

Nuclear power stations are characterized by high initial cost, a low fuel bill, and a great economy of size (see section on the economics of nuclear power). In the hierarchy of units in an electrical network their place, therefore, is as base-load suppliers. This fact sets a natural limit on the fraction of the general demand for electricity that may be satisfied by nuclear energy. For if the nuclear stations were to be operated at low capacity factors, their advantage over fossil stations would disappear and the cost of electricity produced by the network would rise. The exact proportion will depend on the characteristic pattern of the electrical load curve and will vary from country to country. If some form of energy storage (e.g., pumped storage) is included in the network, the nuclear contribution can be increased.

If we now make the unrealistic assumption that all of the baseload of a system will be provided by nuclear stations, we shall obtain an optimistic estimate of the contribution of nuclear energy to the particular country considered. There may well be some situations where such a scenario is valid, but in general there will be several reasons for sharing the base load among several primary sources, and in such cases the nuclear contribution will be lower.

This restriction of nuclear generators to base-load application is not too rigid. The economics in particular situations may be such that some load-following may be acceptable, and the nuclear plants may be allowed to operate part of the time with reduced capacity. Clearly, this will be so as long as the cost of producing electricity, as calculated with the reduced capacity factor, is still lower than that using the alternative fossil fuel. All these considerations lead one to estimate that the global nuclear fraction is approximately equal to the electric base-load fraction. Combining this with the fraction of the total primary energy usage devoted to electricity generation, one can guess at the potential nuclear contribution. For example, if we take 35% as the fraction of primary energy dedicated to electricity production and a nuclear contribution to the latter of, say, 60%, we calculate an overall nuclear share of 21%.

Table 4.2. *Nuclear-generated electricity in some countries, 1985*

	Nuclear electricity (TWh)	Percentage of total electricity
United States	383.7	15.5
France	213.1	64.8
USSR	152.0	10.3
Japan	152.0	22.7
Germany (FRG)	119.8	31.2
Canada	57.1	12.7
Sweden	55.9	42.3
United Kingdom	53.8	19.3
Belgium	32.4	59.8
Spain	26.8	24.0
Industrialized countries	1,313.8	17.9
Developing countries	87.9	4.2
World total	1,401.6	14.9

Source: International Atomic Energy Agency (1986).

From this discussion we can see that nuclear energy, as practiced today, is not the universal answer to all our energy problems, and that even the most aggressive nuclear power policy is unlikely to supply, under the foregoing assumption, more than 20% of the total primary energy needs. Later, we shall discuss ways to raise this contribution closer to 100%.

At the present time the nuclear fraction is well below the limits set in the previous paragraphs. The reasons for this situation will be discussed in the following pages.

In Table 4.2 we show the share of nuclear power in electricity production in 10 countries. These countries account for about 90% of the world's installed nuclear electricity generating capacity and produce about 90% of the nuclear-generated electricity. For most of them the contribution of nuclear power is a smaller fraction of the total electricity generated than one could expect from the previous arguments. There are several reasons for this situation, which we shall discuss shortly. Economic factors undoubtedly play a large role. The general slowdown in the world's economy in the late seventies and early eighties reduced

Table 4.3. *Comparison of electricity generation costs (percentage share of main components)*

Main cost components	Nuclear	Coal	Oil
Capital investment	55–80	25–55	10–25
Fuel	15–30	40–65	70–85
Operation and maintenance	5–15	5–10	5

Source: Bennett (1985).

the demand for electric power to the point that overcapacity occurred. The projections of the utilities were drastically reduced, and plans for the installations of new nuclear stations were canceled or postponed. In addition, the cost of building nuclear stations has escalated faster than the general rate of inflation, thus making it difficult for utilities to raise the required capital.

4.2 Economics of nuclear power

Superficially the generation of electric power from nuclear fission is a very simple process. There is a hot core (not requiring constant fuel injection or air supply), boiling water to raise steam, which is then fed to a conventional power generating train to produce electricity. If this were the whole picture, nuclear power would be exceedingly cheap. In reality, however, the situation is very different. The need to deal with the by-products of the fission reaction (radiation and radioactivity) and to ensure the safety of the workers in the plant and of people living around it dictate many additional features that entail large investments. Thus, the biological shielding, the containment, and the many auxiliary systems needed to ensure safety under all conditions add up to a considerable fraction of the total cost of the plant.

A comparison of electricity generation costs from nuclear and fossil fuels is shown in Table 4.3. Table 4.3 shows that of the three electricity generation systems compared, the nuclear has the highest capital costs and lowest fuel cost, whereas for oil the opposite holds. The final cost of electricity generation in a given situation will depend on many factors, which, taken together, will determine the relative competitiveness of the various systems. For example, high cost of capital will hurt nuclear

Table 4.4. *Cost of construction of hypothetical nuclear and coal-burning stations due to operate in 1995 (in 1982 dollars per kilowatt-electric capacity installed)*

	Nuclear 1 × 1200 MW(e)	Coal 2 × 600 MW(e)
Direct costs		
Land and land rights	5	5
Structures and improvements	190	75
Reactor/boiler plant	240	340
Turbine plant	190	160
Electric and other	155	105
Subtotal direct costs	780	685
Indirect costs		
Engineering and construction services	495	135
Direct plus indirect costs (base)	1,275	820
Owners' costs	125	85
Contingency allowance	210	135
Fore cost (overnight)	1,610	1,040
Escalation (as spent dollars)	1,440	1,155
Interest (as spent dollars)	1,610	755
Plant capital costs at time of commercial operation	4,660	2,950

Key assumptions: Inflation rate, 6% per year; escalation rate of power plant construction cost, 8% per year; average cost of money, 11.9% per year; lead-time, 12 years for nuclear; 8 years for coal.
Sources: DOE (1982); Weinberg et al. (1984).

stations more than oil-burning stations, but the high cost of oil will give the edge to nuclear power. We shall discuss some of these factors more fully in the following paragraphs.

It is of interest to analyze the reasons for the higher capital costs of nuclear stations. In Table 4.4 we present a somewhat more detailed breakdown of the costs of a nuclear-powered electricity generating station, and we compare it with a coal-fired station of the same total power

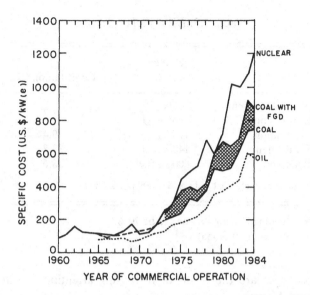

Figure 4.1. Capital costs of nuclear and conventional power stations (mixed years dollars per kW(e); FGD = flue gas desulfurization). (Source: International Atomic Energy Agency, 1982a.)

output, built to today's designs and specifications, to be operational at the same time. Table 4.4 should be regarded only as an example to indicate the major factors contributing to the difference in costs between the construction of future nuclear and coal stations. In an actual case, local conditions and financial terms may produce differences in detail, but it is believed that the main features will remain unaffected.

We see from Table 4.4 that the nuclear plant costs about 60% more than a coal one. We note further that the main difference arises not from the direct costs but from the engineering and financing items. In the early days of nuclear power, reactors were simpler and less encumbered with safety devices, and their costs were correspondingly lower. The cost of reactors installed in the United States in the early seventies averaged $146 to $279 per kilowatt for the various makes. Scaling up to 1983 dollars, these reactors cost $480 to $725 per kilowatt. For comparison, reactors under construction in the early eighties cost between $1100 and $3500 per kilowatt (in 1983 dollars). Conventional coal-burning stations also escalated in cost, as the result of the more stringent environmental requirements, but to a much smaller extent, as Figure 4.1 shows. Most

Table 4.5. *LWR fuel cycle costs*

	Unit cost (U.S. $)	Contribution (%)
Natural uranium	75/kg ⎫	29.2
Conversion	6/kg ⎭	
Fabrication	175/kg	9.4
Enrichment	140/SWU	37.8
Back-end	800/kg fuel	23.6

Total fuel cycle cost = U.S. $ 0.009 per kWh(e)

Reactor load factor assumed to be 70%.
Source: Bennett (1985).

of these changes are the result of increasing attention to safety and environmental issues.

The next major component of the cost of electricity generation (see Table 4.3) is the fuel cycle. The nuclear fuel cycle is far more complex than that of the fossil fuels (see Chapter 3). Recall that it involves all the stages from the mining of the uranium, through its purification, conversion, isotope enrichment, fuel element fabrication, to its ultimate storage, reprocessing, and disposal as active wastes. The cost of all these operations is taken as the cost of the nuclear fuel cycle. The contribution of the components to the total fuel cycle cost of a typical LWR is shown in Table 4.5.

The data of Table 4.5 allow us to make some general observations regarding the sensitivity of the cost of nuclear power to variations in the major cost components of the fuel cycle. We can compute, for example, that an increase in the cost of natural uranium from the reference value of $75 per kilogram to the $130 limit will change the cost of electricity produced by typical LWRs by 2% to 5%. Raising the cost of uranium to the expensive category, $260 per kilogram, will raise the cost of electricity by 11% to 23%. These examples illustrate the point that the cost of uranium is not a major factor in the economics of nuclear power, even in the most demanding strategy (once-through LWR). Other reactor and fuel cycle strategies are still less sensitive to the cost of uranium.

Referring again to Tables 4.3 and 4.4, we note that the most expensive component in producing nuclear electricity is plant construction and that

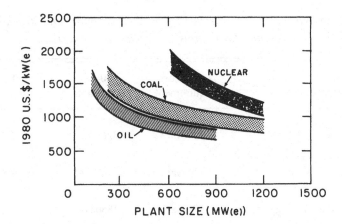

Figure 4.2. Fore costs of nuclear power stations. (Source: International Atomic
Energy Agency 1982a.)

it is heavily influenced by factors depending only weakly on design
power. It follows that there is a considerable economy of size and that
large reactors are to be preferred; once built they had best be operated
as base-load stations. Figure 4.2 shows the fore costs (see Table 4.4 for
definition) of nuclear and fossil stations of various sizes.

Discussions of the absolute costs of nuclear power are confused and
complicated by the different assumptions made regarding parameters
such as the national and utility discount rates, plant size, extent of
environmental regulations, duration of construction, modification of design, ultimate waste disposal, decommissioning, financing, taxation, and
insurance. A detailed discussion of these matters is outside the scope of
this book. At the present time it seems that in general the cost of nuclear
electric power in most countries is close to that produced from coal. It is
cheaper than that generated from the expensive oil of the crisis days and
more expensive than that produced from the cheap oil of the "price-
war" days.

4.3 Availability of nuclear fuel

Up to now, we have been examining the possible share of nuclear power
in electrical networks mainly from the technological and network management aspects. We have not considered the availability of nuclear fuel
to support the various programs.

For conventional power generation systems, the relationship between

Table 4.6. *Nuclear fuel requirements for initial inventory and 30 years of operation for various fuel cycles (in tonnes)*

Fuel cycle and reactor type	Initial core requirement	30-year consumption	30-year production
Once-through LWR Current technology	355 Nat	4260 Nat	5.2 Pu
Once-through HWR Current technology	130 Nat	3655 Nat	9.8 Pu
Pu recycling LWR Improved technology	355 Nat	1850 Nat	2.2 Pu
Pu recycling HWR Current technology	134 Nat	1820 Nat	1.4 Pu
Thorium cycle HTR HEU recycle and makeup	370 Nat 47 Th	1650 Nat +20 Th	~0 Pu
U/Pu cycle FBR Advanced technology	92 Dep	49 Dep +2.6 Pu	10.4 Pu

Note: Based on 1000 MW(e), 70% capacity factor, and 0.2% tails assay. LWR = light-water reactors; HWR = heavy-water reactors; HTR = high-temperature reactors; FBR = fast breeder reactors; HEU = high-enrichment uranium; Dep = depleted uranium; Nat = natural uranium; Th = thorium; Pu = plutonium. *Source:* IAEA (1980a).

the electrical demand and the amount of fossil fuel needed to satisfy it is straightforward. The conversion efficiency is known to within narrow limits, and, with its help, one can calculate readily the future demand for fossil fuels.

In dealing with nuclear power, the situation is more complicated. The translation of the demand for nuclear power into a demand for uranium (or thorium) depends on the types of reactors in the network and on the fuel cycle chosen. In Table 4.6 we present the lifetime (usually taken as 30 years) requirements for uranium of 1000-MW(e) reactors, or their equivalent, of various types, operating with different fuel cycles.

We note from Table 4.6 that current technologies (once-through fuel

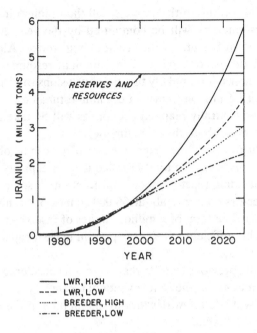

Figure 4.3. Cumulative natural uranium requirements for single strategies (WOCA). (Based on Nuclear Energy Agency/International Atomic Energy Agency 1983.)

cycle) for LWRs and HWRs require that one set aside approximately 4000 tonnes of natural uranium for the lifetime fueling of each 1000-MW(e) reactor built. Reactors of more advanced technology, with pluto-nium recycle, will require about half this amount of uranium, and breed-ers require relatively small amounts. In order to estimate the future needs for nuclear fuel, we must assume a certain scenario for the future mix of reactor types in the national networks. In Figure 4.3 we show such cumulative uranium-use curves based on a number of nuclear power strategies.

We may now compare the cumulative demand for uranium with the estimated reserves. From the discussion of the availability of uranium in the previous chapter, recall that the known reserves and resources at a cost of under $130 per kilogram, excluding the area of centrally planned economies, were estimated at 4.5 million tons. A line at this level was drawn in Figure 4.3. It will be seen readily that a strategy involving only LWRs in the single-pass fuel cycle calls for the largest commitment of uranium. According to this strategy, all the uranium in the reserves and resources category will be committed by the year 2020 – that is, toward the end of the life of the reactors built today. Although these numbers exclude the reactors and the uranium resources of centrally planned economies, it is unlikely that the conclusions will change materially. The extra 1.5 million tonnes of uranium estimated to be present in the area of the centrally planned economies will be taken up by their nuclear program on about the same time scale.

Another illustration of the magnitudes involved can be obtained from the following numbers. The total installed nuclear capacity in the world at present (including reactors under construction) is some 400 GW(e), and these reactors consume about 56,000 tonnes of uranium per year (see Table 4.6). The total of 6 million tonnes of reserves and resources will therefore be sufficient to keep the present nuclear capacity going for just over 100 years.

We saw in the previous chapter that there is a considerable amount of additional uranium in the unconventional category (e.g., in phosphates). This will add a few decades to the lifetime of the LWRs using the single-pass fuel cycle.

Because the cost of nuclear electricity is not a very sensitive function of the fuel cost, the distinction between economical and uneconomical uranium is not sharp and depends on local circumstances. The limit of $130 per kilogram for the cost of uranium is also somewhat arbitrary. If this limit is raised, additional resources of uranium become available, and the lifetime of all strategies is extended.

From Table 4.6 and Figure 4.3 we see that other reactor and fuel cycle choices lead to smaller uranium requirements. Fuel cycles involving plutonium recycling and especially breeder reactors are far more economical in their uranium demands than are single-pass LWRs. Extensive studies were devoted to possible mixed strategies according to which types of reactors are incorporated in certain proportions in the networks. Mixed strategies that include a high proportion of breeder reactors conserved the most nuclear fuel (see Table 4.6).

It is obvious that the rate of breeder deployment is limited by the availability of plutonium for the initial loading and first few reloads. The plutonium can only be obtained through the operation, over a period of time, of LWRs and HWRs. A mixed strategy for several decades is therefore unavoidable.

We note (see Table 4.6) that the amount of uranium needed to fuel breeder reactors is very small (49 tonnes for 30-year operation of a 1000-MW(e) unit), and, moreover, it can be depleted uranium. The amount of this material available from the tailings of the isotope-enrichment plants and from the spent fuel elements of reactors will be very large, within a few percent of the total natural uranium introduced into the fuel cycle. This means that as far as breeder technology goes, we can consider practically all the initial uranium reserves and resources as still available today. To illustrate the implication of these numbers, we can calculate that the known reserves and resources of conventional uranium will be sufficient to supply all the world's 1985 electricity needs (9422 TWh) for 1800 years if it were all produced by breeder reactors. If the additional nonconventional uranium resources are taken into account, this period can be more than doubled. Further, since the fuel component of the cost of electricity produced by breeder reactors is very small, much more expensive uranium can be considered as economically acceptable, and this opens up the vast resources of the rocks ("rock burning") and the oceans. From the data presented in Chapter 3 it can be calculated readily that if all the uranium present in the ocean were burnt in breeder or advanced converters, about 200 million exajoules of energy could be released. If we include the reserves of thorium that can also be tapped in advanced reactor concepts, we see that fission energy, via breeders and other advanced reactors, offers an essentially unlimited source of energy. However, at this writing there is great opposition to breeder reactors, and their extensive deployment is not in sight.

Recall that any strategy involving other than single-pass LWRs or HWRs calls for reprocessing and plutonium recycling. If this is unacceptable, the presently known reserves and resources of cheap (i.e., at less than \$130 per kilogram) uranium will last to the middle of the next century.

In once-through strategies (i.e., no reprocessing) large amounts of plutonium and reusable uranium accumulate in the form of spent fuel elements. The temptation will be great, particularly when the cheaper uranium begins to run out sometime in the beginning of the next century, to reprocess the spent fuel and use the fissile and fertile components. It is likely, therefore, that long before the reserves and resources (see Table 3.3) run out, some mixed strategies involving civilian reprocessing, plutonium recycling, and breeder reactors will develop.

So far, we have discussed this issue from a global point of view. In the absence of a world energy policy, actual decisions will be made by nations in accordance with their particular situation. Those that have ample and assured uranium supplies will probably defer the recycling of plutonium and the introduction of breeder reactors. Those nations that possess little or no cheap uranium will opt for the early introduction of fuel-efficient nuclear cycles and breeder reactors. Indeed, we see this trend already in operation. We note that the United States and Canada have no civilian reprocessing and no serious breeder development programs. On the other hand, Japan and some European nations have vigorous breeder development programs and civilian reprocessing facilities.

The discussion so far shows that in nuclear energy we have the first source that is potentially perpetual, is based on a proven technology, and is economically sound. In spite of all this, the future role of nuclear energy is still shrouded with doubts. The reasons for this situation and the actions needed to correct it are discussed in the following section.

4.4 Opposition to nuclear power

There is no doubt that the introduction of civilian nuclear power has been delayed, and even prevented in some countries, as the result of growing public opposition. This opposition is the outcome of many and complex factors – some physical, others psychological – that act directly and indirectly against nuclear power. These factors arise from the association in people's minds of nuclear power with nuclear weapons, from their anxiety about reactor safety, and from concern about the environment. Indirectly, the effect of this opposition has been to escalate the cost of construction of nuclear plants to the point that their economic advantage is disappearing. This comes about as a consequence of drawn-out licensing procedures, changes in design, redundant safety features, and long construction times. These factors, coupled with the public opprobrium, make the installation of a nuclear power station an unattractive proposition to a public utility. The foregoing relates primarily to the situation in democracies.

The question of high-level radioactive wastes is often raised in connection with the licensing of new reactor construction. Fuel reprocessing, when pursued, takes place in special facilities that serve the entire national nuclear program and are not related to any specific reactor. Therefore, when this issue is raised in connection with the approval of a specific reactor construction, it is only an expression of the opposition to the nuclear program as a whole, and is really irrelevant to the particular site and its neighbors.

Because of the serious implications of these issues to the development of nuclear power and the inherent importance of the basic questions raised by them, we shall discuss briefly some of the main problems.

4.4.1 Proliferation and diversion

There is no doubt that some of the technologies involved in the fuel cycle can be turned to weapons production. Thus, the possession of an isotope-enrichment technology can enable a nation to produce weapons-grade uranium. Similarly, the possession of a reprocessing plant can enable a nation to separate weapons-grade plutonium from irradiated uranium. Because the amount of fissile elements in circulation in the fuel cycle from even a moderate nuclear power program is very large compared with the needs of a modest weapons program, the diversion of even a small percentage of the inventory can enable a clandestine military program to flourish. This is the reason for the severe controls imposed by the Nonproliferation Treaty (NPT) on all nuclear facilities and particularly those involved in the fuel cycle. How effective these controls are is a moot point.

The possession of a nuclear power plant does not in itself enable a nation to produce nuclear weapons. For example, a nation that chooses to operate a light-water nuclear power plant on the basis of imported enriched fuel and stores its spent fuel elements or sends them abroad to approved reprocessing plants does not acquire a nuclear weapons capability. The capability of producing nuclear weapons becomes apparent when a nation acquires either isotope-enrichment or fuel reprocessing facilities. Such situations are beginning to develop outside the nuclear nations (defined by the NPT as those nations that manufactured and exploded a nuclear weapon or other nuclear explosive device prior to January 1, 1967). Because the breeder technology includes reprocessing as an indispensable part of the nuclear fuel cycle, it is clear why this particular route to nuclear power is so vehemently objected to.

Up to now we have been considering the case of a government trying to circumvent the nonproliferation agreement. It is doubtful whether a government intent on implementing a clandestine nuclear weapons program will choose the power reactor route – it has so many other options that are easier to conceal. But there is another possibility that causes anxiety to both nuclear states and nonnuclear states, and it is the danger that fissile material will fall into the hands of terrorist organizations. The amounts of material involved in this danger are even smaller than those needed for national clandestine programs. One device, even if crude, in the wrong hands can cause an untold amount of trouble. Hence, any facility where high-grade fissile material is present must be guarded by the most sophisticated means available. It is this aspect, involving extensive participation of a central authority in a purely civilian undertaking, that is also found objectionable in many quarters.

Considerable ingenuity is now being devoted to devising nuclear fuel cycles that inherently offer more safeguards against diversions. Suggested stratagems include irradiating the fuel, adding radioactive isotopes to it, or leaving some of the fission products in it to make the handling of the material, without the proper special equipment, more difficult and dangerous. Another line of thought is to "denature" the fissile material by adding to it a nonfissile isotope and thus making it unsuitable for weapons manufacture. All these measures complicate matters and add to the cost of the fuel cycle. So far none of these ideas has been adopted.

4.4.2 Reactor safety

The issue of reactor safety figures prominently in all discussions regarding nuclear power. This is natural, considering the basic difference between accidents involving conventional power plants and those involving nuclear reactors. In the former, only the plant's personnel are at risk; in the latter, the public outside the plant's perimeter, or even outside the national borders, is also endangered. It is recognized that there is a great difference in the perception of risk by individuals if it is taken voluntarily (driving a car or seeking employment in a nuclear facility) rather than imposed. Because of this difference, there is great sensitivity to any potential accidents involving nuclear power reactors, even if they have a very low probability. This concern, which is often fanned by those who oppose nuclear power for other reasons, causes local resistance to the erection of power stations and may prolong licensing procedures. On the positive side it also forces the utilities and suppliers to pay much greater attention to safety measures, proper design, good workmanship, and effective quality control.

There has been a drive in recent years to increase the sophistication of the safety measures and their redundancy. It is possible that a condition of diminishing returns has already been reached where an increase in the safety measures, each with its own failure rate, does not lead to an increase in the reliability of the system as a whole.

The type of accident most feared in the nuclear power industry is that caused by the loss of coolant. If, for one of many possible reasons, the flow of coolant is stopped while the reactor is operating, a rapid rise in temperature of the core will occur, even if the reactor were scrammed immediately, because of the large heat evolution from the decaying fission products in the fuel elements. Some idea of the magnitude of this heat evolution can be obtained from the following numbers. Two hours after the shutdown of a 1000-MW(e) reactor that has been operating for only a month, the decay heat is 30 MW(th) – enough to boil away almost 50 tons of water per day. Even one week after the shutdown of a similar reactor that has been operating for a year, the heat evolution is sufficient to boil away almost 15 tonnes of water per hour. It is clear that with these rates of heat evolution, and without special provisions for emergency cooling, the core will soon melt, leading to the most dreaded accident. For this reason the design of all reactors includes elaborate emergency cooling systems, which should provide protection against the effects of loss of coolant circulation. It is only when these systems fail or are circumvented and the inventory of the coolant is lost that the reactor core can dry out and melt.

A meltdown not only permanently destroys the reactor, but it also releases a great deal of the radioactivity contained in the fuel. Nor does the process end there; heat continues to be generated within the molten mass, which may proceed to melt its way through the containment and ultimately escape into the environment (the facetiously called China syndrome). Needless to say, the most elaborate precautions are taken in the design of reactors to avoid such a situation and minimize its consequences if it ever should occur.

The nuclear industry has accumulated several thousands of reactor-years of commercial operation, with only a few accidents. This would be considered a good record if it were not for the fact that one serious accident has such a great social, economic, and environmental impact that it can wipe out much of the advantage of nuclear power. Of the two major accidents that have occurred so far, Three Mile Island and Chernobyl, only the first has relevance to the question of power reactor safety. The reason is that, whereas the Three Mile Island accident involved a power reactor in routine operation, the Chernobyl reactor was being used at the time of the accident as an experimental facility operated well outside its design conditions and in an unauthorized manner. Both involved a partial meltdown of the core, and a release of the radioactivity contained in it. The different outcomes of the two accidents is a consequence of differences in the design of the two reactors and the way they were being operated at the time. In the Three Mile Island reactor the design seems to have been adequate to contain the accident, and no fatalities or damage to the environment ensued. Chernobyl clearly was not designed to deal with the type of experiments being carried out. The two accidents occurred in technologically advanced countries, and both can be traced to human error. The lesson from these accidents is clearly that operators are fallible and that the safety systems, as designed, could fail or be circumvented. Another conclusion is that power reactors that are *inherently safe* must be designed – that is, reactors that do not depend for their safety on the proper functioning of instruments or operators.

In recent years (even before Chernobyl) there has been considerable interest in and discussion of the feasibility of inherently safe reactors, but no development program has been initiated. There is an increasing awareness that unless a satisfactory safe design is developed, the future of nuclear power is in doubt. No amount of explanation of the two major accidents will increase public acceptance of nuclear power. I suspect that an actual demonstration of an inherently safe reactor will be needed to restore public confidence in this technology. After all, if a reactor really is inherently safe, there should be no difficulty in publicly demonstrating this fact. Success in these endeavors will materially affect the share of nuclear power in the energy economy (see Chapters 3 and 6).

Perhaps it is not too optimistic to hope that the economic advantage of nuclear energy and its promise of perpetual availability will encourage the development of inherently safe reactors and the discovery of an acceptable long-term solution to the nuclear waste problem. In a sense the process of development of nuclear power was stopped too soon, and its commercial application started too early; it is essential to go back and complete the former development, because few other long-range alternatives exist, and they can be exploited only after much research and development.

Additional reading

Häfele, W., Harms, A. A., Bauer, G. S., McDonald, A., (1983). *Nuclear Technologies in a Sustainable Energy System.* Springer-Verlag, New York.
Nuclear Energy and Its Fuel Cycle. OECD/NEA (1982). OECD, Paris.
    (1985). *The Economics of the Nuclear Fuel Cycle.* OECD, Paris.
Weinberg, Alvin M. et al. (1985). *The Second Nuclear Era: A New Start for Nuclear Power.* Institute for Energy Analysis, Oak Ridge, Tennessee. Praeger, New York.

5

Renewable energy resources

The fascination of renewable energy resources derives from the desire to break away from the present almost total dependence on fuels of finite lifetime and from the appreciation that ultimately mankind will have to rely largely on some inexhaustible sources. Obviously, any energy source that is renewed on a short time scale is effectively inexhaustible. Unfortunately, not many such sources are of sufficient magnitude to supply an appreciable fraction of mankind's needs.

Until a few hundred years ago mankind did depend entirely on renewable energy sources, but the large and rapid industrialization that started in the last century is based mostly on the exploitation of nonrenewable resources. This chapter differs from previous ones in that the discussion of the current situation will be brief, and that of future prospects will be longer. This reflects the present status of the share of renewable sources in the energy economy.

It will be appreciated that, apart from geothermal and tidal energies, all other renewable energy sources on earth are essentially derived from the sun. We shall discuss all of these in this chapter, starting with the various forms of solar energy.

The amount of energy reaching the earth from the sun is truly prodigious. Each square meter facing the sun at the distance of the earth's orbit receives a flux of 1372 W. The total energy intercepted by the earth is about 5.6 million EJ/yr. Of this 1.7 million EJ/yr are reflected or scattered back into space, leaving 3.9 million EJ/yr to be absorbed by the atmosphere and the surface of the earth. This energy is some 12,000 times the total present world consumption. Some of this is naturally converted by geophysical processes to the energy content of the winds,

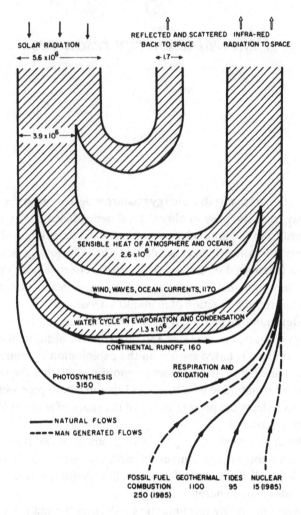

Figure 5.1. Energy balance of the earth (EJ/year).

waves, water running off the continents, and by photochemical processes to biomass. These become, therefore, potential sources of useful renewable energy (see Figure 5.1). But the great bulk of the solar radiation reaching the surface of the earth is unharvested. It is quite natural to ask whether a small fraction of this huge amount can be harnessed and so satisfy all future human needs.

The total solar energy absorbed by the land mass of the earth is approximately 790,000 EJ/yr, but obviously this cannot be regarded as anywhere near the resource magnitude because only a tiny fraction of the surface of the earth can be set aside for energy collection. The solar energy reaching the major hot deserts of the world is estimated at 93,000 EJ (Bockris 1980). This is still about 300 times the present world energy demand. It follows that a few percent of the world's deserts, if assigned to solar energy harvesting, could permanently satisfy all mankind's future energy needs [taken as 1100 to 1500 EJ per year (see Chapter 6)].

Although the total amount of solar energy seems to be quite large compared with human needs, we should not assume that a great fraction of it may be diverted to satisfying those needs. This is because the world's climate, controlled as it is by oceanic and atmospheric circulations, is highly sensitive to the delicate balance between various energy inflows and losses. Although the collection of large amounts of solar energy and its use elsewhere does not represent any change in the total energy balance of the earth, it does involve a change in its distribution. Even relatively small perturbations in balance and distribution, caused by human activity, may lead to major climatic changes and environmental disturbances. It is presently impossible to set precise environmental limits on the maximum permissible solar energy utilization, so we shall take the magnitude of the direct solar energy resource as 93,000 EJ per year, with the caveat that only a small fraction of it, depending on technological, environmental, and economic factors, will actually be used. We have already encountered a similar problem, where the environmental effects of carbon dioxide emissions to the atmosphere may limit the use of fossil-fuel resources (see Section 2.5).

We shall first discuss the possibilities open for utilizing some of the solar energy directly reaching the surface of the earth. Only little use can be made of the direct, unchanged solar energy. It is generally necessary to convert solar energy to some form of secondary energy before it becomes useful to man. We shall now discuss the conversion processes available or under development. Later in this chapter we shall deal with ways of utilizing part of the solar energy that has naturally been converted to other forms.

Some well-known characteristics of solar energy are worth recalling because of their profound effect on the future potential of this resource.

Like the other primary energy sources, solar energy is not uniformly distributed on the surface of the earth. The regions of high availability are limited to certain zones, which frequently do not coincide with the areas of high energy demand. Figure 5.2 shows the regions of high insolation.

The difference between fossil-energy resources and solar energy is that the former can be readily transported (by rail, ship, or pipeline) and stored, whereas the latter cannot be delivered to a general user at any appreciable distance from the point of collection unless it is first converted to another form. At the present time the only secondary energy forms available for the long-distance transport of solar energy are electricity or biomass and derived products. Apart from the very small fraction of biomass energy products exploited presently, there are no methods known today for the transmission of solar energy to the distant nonelectric end-user. Finding a solution to this problem is one of the goals of solar energy research. If no solution is found, the applicability of this resource will be limited to unique local situations or to being a component of the electricity generation system.

Another limitation of solar energy is that it is intermittent. The sun rises and sets at predicted times, and passing clouds can introduce an unpredictable interruption of the radiation. Unless the user can tolerate this, some type of intermediate storage is essential. Converting solar energy to electricity does not help in this problem, unless its contribution is very small compared with the network's size, or unless the electricity network possesses an appreciable pumped-storage capacity. Conversion of solar energy to chemical energy (gases or liquids) could be a possible solution to the storage and transport problems.

5.1 Solar thermal energy

Perhaps the most common and simplest use of solar energy is as heat. In thermodynamic terms, converting solar energy to heat greatly degrades the value of the energy. The radiation arriving from the sun represents a source at about 6000 K, and converting it to low-grade heat means a great loss of potential efficiency (see Section 1.4). The lower the temperature level of the absorbed solar radiation, the smaller is the general thermodynamic efficiency of any further conversion process. We shall discuss the different solar thermal technologies according to the temperature level of the heat produced by the capture of the solar radiation.

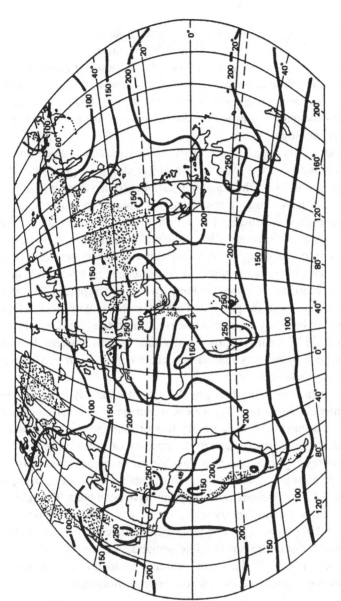

Figure 5.2. Insolation map of the world (W/m^2 averaged over 24 h). (Source: Kreith and Kreider 1978; copyright Hemisphere Publishing Corp.)

5.1.1 Low-temperature solar heat
5.1.1.1 Domestic hot water

By far the simplest and most common application is the capture of solar heat at temperatures below the boiling point of water (say around 70°–80°C). The technology for doing this is well known and reliable. Rooftop collectors for providing domestic hot water are a good example of this technology.

In spite of its apparent simplicity, the technology is sometimes quite sophisticated. For example, in the better designs, spectrally selective coatings are used on the collecting surfaces to increase their efficiency. These coatings are highly absorbent (i.e., black) to visible solar radiation but emit very little heat radiation in the infrared. Other ingenious devices aimed at increasing the collecting efficiency are also often incorporated into the designs. In warm climates the system can be quite simple, with no moving parts. In colder climates where freezing may occur, a more complicated system, involving a separate heat-transfer fluid, must be used. In spite of the great simplicity of this technology and its wide applicability, its actual contribution to the global energy supply is minuscule. Even in a sunny country such as Israel, where the installation of such systems is mandatory, their contribution to the primary energy supply is only about 2%.

If we turn now to another possible use of relatively low-temperature heat, namely space heating, we encounter an additional complication. For such a purpose it is necessary to store large amounts of heat captured during the day for use in the evening or, even more difficult, to store energy captured on sunny days for use on cloudy ones or from one season to another. Much work on this problem has, and is, being done, and a number of demonstration systems have been built. The subject has not yet reached a satisfactory technological and economical status.

The global potential for low-temperature solar thermal energy is enormous and far outstrips any possible demand. A few tens of thousands of square kilometers of the earth's surface could provide all mankind's need for low-grade heat, even with present technologies.

5.1.1.2 Solar ponds

The efficiency of converting heat energy into electricity drops drastically as the source temperature is decreased (see Section 1.4). Thus, the maximal thermodynamic (Carnot) efficiency possible with a source temperature of 80°C and a sink at 30°C is 14%, and a practical 10% would be considered a great achievement. For this reason, low-temperature solar heat is not an attractive route to electricity production. Nevertheless, proposals exist for utilizing low-temperature solar heat, based on systems where the advantage of the low cost of the energy-collecting components outweighs the disadvantage of the low generation efficiency.

Two examples of low-temperature solar electricity production will be described. The first is the *solar pond* (EPRI 1985a). This process relies on maintaining a gradient of salt concentration in a water pond so that a high-salinity layer is present near the bottom and a layer of almost fresh water near the top. (See Figure 5.3.) A properly adjusted concentration gradient will prevent thermal convection, and, once this is eliminated, the lower layers of the pond will heat on absorption of solar radiation. The hot water from the bottom of the pond, at around 80°C, can be drawn off and used to boil a carefully selected organic fluid. The vapors of this fluid drive a specially designed low-temperature turbine to generate electricity.

A most important feature of this process is its heat storage ability. The large mass of the pond water can provide seasonal storage capacity, and thus the utilization of its energy is completely uncoupled from the solar radiation. This makes the source flexible and suitable for intermittent and peaking supplies. It is similar to hydroelectric power in this respect and in its broad economic features. It is capital intensive, involves extensive civil engineering work, the fuel costs nothing, and the operating and maintenance costs are low. The solar pond concept has been under development in Israel for the last 25 years, and a succession of larger and larger ponds has been built and operated. The largest pond in operation in Israel, and probably in the world (250,000 m²) is on the shores of the Dead Sea (Bronicki 1984). It is fitted with a 5-MW turbine connected to the national grid.

Although the concept in its present form requires suitable topography for large-scale implementation, a survey has shown that many locations around the world have the right conditions. To illustrate the potential of this technology, we can readily calculate that even with the low overall conversion efficiency demonstrated so far (around 1%), 0.3% of the total land mass surface, if covered with solar ponds, could be sufficient to supply all the present electricity needs of the world.

A variant of the solar pond, the floating pond, is now under development. The floating pond would make it possible to use protected bodies of water such as lakes and bays for solar pond construction. This would

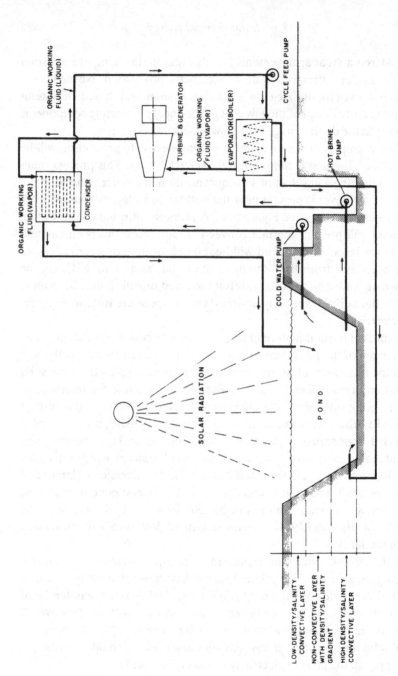

Figure 5.3. Schematic of a solar pond. (Source: Bronicki 1984.)

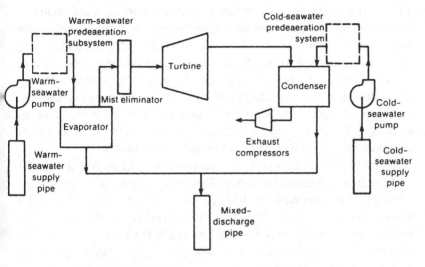

Figure 5.4. Schematic of an OTEC open-cycle (Claude) system. (Source: Kreith and Bharathan 1988.)

obviate the expense of large earth-moving projects and would increase the number of locations where such technology might be installed. However, the role of this new technology in global renewable energy supplies is still not clear.

5.1.1.3 Ocean thermal gradient
Another approach to the use of cheap low-level solar heat for electricity production is based on naturally occurring temperature differences between the surface of the ocean and deep water. In several places around the world large differences, which may even reach 24°C, occur. With such temperature differences, the efficiency of electricity generation will be even lower than in the solar pond (a theoretical maximum of 8% and a practical maximum probably not greater than 4%). Nevertheless, the enormous magnitude of the resource makes the approach worth studying (Isaacs, and Schmitt 1980; Cohen 1980). Present projects, which go under the name of Ocean Thermal Electricity Conversion (OTEC), are based on two concepts: the open cycle and the closed cycle.

In the open (or direct) cycle, first proposed and tried by Claude off Cuba in 1926 (Claude 1930), warm surface seawater is allowed to boil under its own vapor pressure and the vapors, after passing through a turbine, are condensed by cold water pumped from the ocean depths. (See Figure 5.4.) Claude succeeded in generating a gross output of 22 kW, but no net power was obtained. In its simplest form the cycle can involve the condensation of the vapor coming through the turbine by direct contact with the seawater, the combined condensate and cooling flows being returned to sea. In another scheme that may have an economic advantage in certain localities, the vapors are condensed in a conventional condenser (flat plate or tube-and-shell types) and the fresh, desalinated, water is sold as an additional product of the plant. In all these schemes the low pressure in the system (less than 25 mbar) dictates very large volumes and enormous turbine blades.

In the closed cycle, the warm seawater is used to boil a suitable working fluid (ammonia or freon). The vapors of this fluid then drive a turbine and are condensed with the help of cold water pumped from the deep layers of the ocean. This concept was tried out in Hawaii in 1979, with ammonia as the working fluid. A 50-kW (gross) mini-OTEC was mounted on a barge moored off Keahole Point. It generated a net power of 15 kW. A larger experiment was carried out in Japan in 1982, where a 100-kW (gross) unit, using freon as the working fluid, was tested off the island of Nauru. A net power production of 35 kW was demonstrated. The low inherent efficiency of the process means that very large volumes of water have to be pumped and very large heat exchangers must be employed. The surfaces of these heat exchangers must be kept scrupulously clean and free from fouling to avoid degradation of the heat-transfer capability. These difficulties put the closed-cycle concept at a disadvantage when compared with the open, direct cycle. The main difficulty with the latter concept, that of the very large turbines required even for modest power output, led to the development of another variant, the mist-lift open cycle. In this system the warm water is allowed to boil under vacuum in such a way that a large proportion of mist is carried by the vapors. The vapors ascend a tall "chimney" where they are condensed by the cold water. The combined condensate and cooling water is then allowed to fall back to the ocean through a hydraulic turbine of normal dimensions. This concept has not been tested yet.

One of the main engineering problems encountered was the design of the cold-water pipe. This large-diameter conduit has to extend to great depths and withstand the harsh conditions of the open sea.

In addition to the purely engineering problems, there are some conceptual difficulties that may limit the use of this resource to well below the magnitude that one would estimate on the basis of the area of the tropical and subtropical seas.

If only shore-based stations are considered, then the choice of locations is severely limited to those places where cold deep water is found close to the coast. A side benefit of shore-based stations is the nutrient-rich bottom water that will be brought to the surface. This could provide a base for a lucrative aquaculture industry.

If floating stations are included, then there is a host of new difficulties in addition to the obvious ones of operating large plants for many years at sea. There is the question of what is to be done with the electric power generated far out at sea, because the transmission of large amounts of power over long distance underwater presents a major engineering challenge. The need for this latter technology could be avoided if an oceangoing OTEC plant were integrated with a large electric power-consuming industry that was also floating nearby or on the same platform.

There are several possibilities for such integration. For example, hydrogen can be made by the electrolysis of water and piped to the shore or used in situ to produce ammonia, which is then shipped to market. Another possibility is the recovery of aluminum from purified ores delivered to the plant by sea. Such integration limits the large-scale deployment of OTEC technology and constrains it to the market for the commodities produced.

We may summarize the discussion by pointing out that because the application of OTEC will be limited to a fraction of the electricity sector, its global impact will also be limited by and linked to the growth of this sector.

The discussion of the maximum possible global potential of the OTEC technology is difficult even if one is prepared to assume that all the engineering and economic problems have been solved. The open questions relate to the impact of large-scale deployment of OTEC stations on the delicate thermal balance of the earth. Solar radiation is absorbed by the tropical and subtropical seas, and the heat thus generated causes ocean currents and atmospheric circulation that distribute the thermal energy to the temperate and polar regions. The world's climate and weather patterns are quite sensitive to this energy transport. The Gulf Stream and its importance to the European climate and the El Nino and its effect on Peru are both well known.

The operation of OTEC stations, if carried out on a large scale, would have the effect of cooling the surface of the ocean and thus affecting the oceanic and atmospheric circulation. We cannot estimate these effects precisely, but some feeling for the quantities involved may be obtained from the following considerations. The total amount of solar energy absorbed by the oceans is about 100,000 EJ/yr. Most of this energy is transferred to the atmosphere, used to evaporate water, or reradiated into space. About 6 EJ drives the ocean currents and controls the climate. Any interference with this small amount could upset the system's balance and drastically affect the climate. But how much interference is tolerable? Without reliable models connecting the many complex phenomena, it is difficult to make a realistic estimate. Wick and Schmitt (1977) suggested that it might be possible to intercept 1500 EJ of the incoming solar radiation and use it for power production. With a conversion efficiency of 3%, this would lead to 45 EJ of electricity, just about the expected world demand in 1995. Whether it is permissible to draw off a few percent of the surface energy of the ocean without leading to major climatic changes is an open question.

5.1.2 High-temperature solar thermal

We have seen that the global applicability of low-temperature solar heat is limited by the relatively small direct demand for heat of this quality and by the low thermodynamic efficiency of its further conversion to electricity. A much greater potential exists for high-temperature solar thermal energy in such fields as industrial process-heat and the generation of electricity. The major development effort so far has been directed to the generation of electricity.

In order to obtain high temperatures, some degree of concentration of the solar radiation is necessary. A variety of concentrating devices have been developed and tested over the years. In all but the low-concentration devices some means of following the sun's apparent motion across the sky must be provided. The degree of sophistication of such tracking devices depends on the degree of concentration desired, which in turn depends on the maximum temperature to be achieved. Thus, if temperatures only slightly above 100°C are needed (as, for example, for air conditioning using absorption cycle refrigeration), a nontracking concentrator may be sufficient. For temperatures up to 300°C, which covers much of the process-heat needs of industry, tracking mechanisms are necessary, although they can be relatively simple. For the highest temperature range, extending above 500°C, which covers the needs of industry and electric power generation, more sophisticated mechanisms are needed to control the concentrating elements.

The requirement of tracking mechanisms increases the complication and cost of high-temperature solar devices and has been one of the main impediments to their commercial application. But recent developments in computers and microelectronics have changed the situation, and now the provision of precise and reliable tracking for even huge arrays is relatively simple.

Demonstration plants have been constructed throughout the world. Some of the smaller ones are devoted to process-heat studies, and the largest are dedicated to electricity production.

The concentrating devices studied include parabolic troughs, which give a line focus, dishes, which give a focal spot, and large arrays of focusing mirrors, which give a composite image disk. Some special shapes and designs for low concentration (2–3 times the solar intensity) have also been proposed and tested.

Regardless of the particular technology used, if large amounts of solar energy are to be collected, extensive reflecting surfaces must be provided. Practical considerations limit the physical size of the individual concentrating element. It follows that the large total collecting surface required for any commercial system must be broken down into many smaller units. If the system is based on parabolic dishes or troughs, the output of the many units must be connected in parallel to achieve the total power specified. Such a system is known as a *distributed collector system,* and the field accommodating the individual units is sometimes called the *farm.*

In a dish farm there are two ways to connect the individual collectors in parallel: Each dish may produce the final product, electricity, and the outputs are then combined; or each dish may produce, as an intermediate output, a stream of a hot heat-transfer medium that, after combining with other streams, is used to generate steam and electricity in a central plant. Parabolic troughs are always combined through an intermediate heat-transfer medium, usually oil.

The long connecting lines of distributed systems are a source of energy loss and complexity.

The central receiver technology avoids this problem by combining the reflected light from many collectors onto a central receiver where water or some other heat-transfer fluid is heated.

Parabolic troughs. The parabolic collector, as the name implies, consists of a long, reflecting trough of parabolic shape with a tube, through which flows the heat-transfer medium, mounted at the focal line. (See

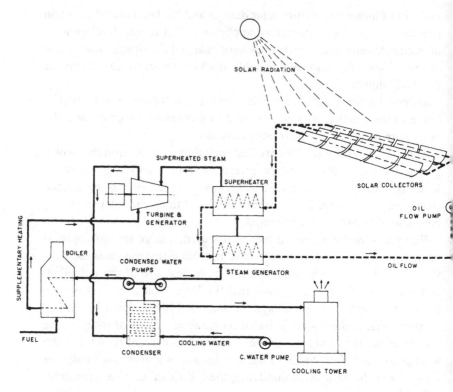

Figure 5.5. A parabolic trough collector system. (Source: Jaffe, Friedlander, and
Kerney 1987.)

Figure 5.5.) The trough can be constructed to run in the east–west
direction or, less frequently, in the north–south direction, and it tracks
the sun in the orthogonal direction. The need for tracking along one axis
only is a simplifying feature of this design. In the best designs radiation
losses are reduced by coating the receiver tubes with a selective surface
and enclosing them in insulating glass sleeves.

The largest parabolic trough plant operating today is the 103.8-
MW(e) facility built by LUZ in California and connected to the Southern California Edison network (SEGS 1 of 13.8 MW(e), and SEGS II,
III, and IV of 30 MW(e) each; see Jaffe, Friedlander, and Kerney 1987).
The plants have numerous collector assemblies, each containing eight
modules and driven by a single tracking mechanism. The early, first-
generation modules consisted of a parabolic trough 6.4 m long and 2.6 m
high made of back-silvered glass. The present, second-generation mod-

ules are about twice as high and have about double the aperture area. The receiver is a tube coated with a selective surface material and enclosed in an evacuated tube. The heat-transfer fluid flowing through the receiver is oil.

Dish collectors. In the dish concept the collector is shaped like a paraboloid reflector. It is usually constructed from many reflecting facets assembled on a steerable frame and adjusted to give "point" focus. (The focus is really a disk corresponding to the sun's image.) The dish has to be steered along two axes to maintain the sun's image on the receiver, which is mounted at the focal point. The receiver may be a complete electricity generating unit, as for example a Stirling engine in a recent design, or a device for heating a heat-transfer fluid. Numerous designs have been built and tested.

The practical limit to the diameter of a dish collector is probably around 15 m, which can yield between 40 and 50 kW of electricity. For larger power it is necessary to connect several dishes in parallel. The largest dish farm, built by LaJet Co., is near Warner Springs, California. Each of seven hundred 7-m-diameter dishes generates steam, which is then combined and used to produce 4.9 MW of electricity at peak. Each dish has an effective aperture of 43.2 m^2 and consists of 24 reflecting facets made of aluminized Mylar vacuum-stretched over circular frames. The plant is connected to the San Diego Gas & Electric Company's network.

A smaller plant, generating 400 kW of electricity, 677 kg/h of process steam, and 3250 MJ/h of refrigeration, has been operating for some time in Shenandoah, Georgia. It consists of 114 parabolic dish modules 7 m in diameter, each heating a heat-transfer fluid. The hot liquid from all the dishes, which are connected in parallel, is combined and used to drive the combined electricity, process-heat, and refrigeration cycle.

Central receivers. Driven by the impossibility of building the very large dish that would be needed to supply any commercial amount of power, engineers evolved the concept of breaking up the large reflecting surface required into many manageable reflectors, each aimed at the same point. Each reflector, which may be up to 10 by 10 m, is itself made up of many smaller mirror facets of the correct curvature and mounted on a two-axis steerable frame. The whole unit is known as a *heliostat.* A commercial power station could have many thousands of heliostats, all reflecting the solar radiation onto a single receiver – hence the name

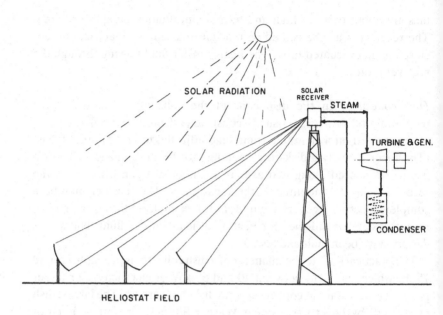

HELIOSTAT FIELD

Figure 5.6. Schematic of a central solar receiver plant.

central receiver for this technology. Simple optical considerations require that the receiver be located at an optimal height above the plane of the heliostat field – hence the popular name *solar tower* for such power plants. (See Figure 5.6.) Various designs for the receiver and power generating loops have been built and tested, using a number of heat-transfer fluids. The latter include, in addition to water, molten salts and liquid sodium.

Demonstration plants for electricity generation based on the central receiver concept have been built and operated to gather economic and technical data. The largest, Solar One at Barstow, California, is designed to supply 10 MW of electricity for 8 h on a summer's day to the Southern California Edison network. It consists of a field of 1818 heliostats, each of 40 m² reflecting area. The reflected solar radiation is directed to a cylindrical receiver, 14 m high and 7 m in diameter, situated atop a 91-m-high tower. Water is boiled in the receiver, and the steam is led either to a turbine or to a thermal storage unit.

Smaller plants have been built in other parts of the world. A 2-MW(e) plant (THEMIS) in the French Pyrenees differs from others in that it uses molten salt (a mixture of nitrates) as the heat-transfer fluid and heat-storage medium. The reflecting field consists of 200 heliostats of 54 m² each that focus the radiation into a cavity receiver situated on top of an 80-m-high tower. This plant operated for a number of years and supplied electricity to the EDF (Électricité de France) network (see Kreith and Meyer 1984).

The potential of high-temperature solar thermal technologies. The various high-temperature solar technologies may be considered together when one tries to assess their maximum potential. This can be done because they are similar in their thermodynamic efficiencies and land requirements. If we assume further that 5% of the solar energy falling on the plant area can be converted to electricity, we can calculate that about 1/1500 of the total land mass of the earth, if devoted to solar power generation by these technologies, would suffice to provide all the present world electricity demand.

Inasmuch as these technologies are not limited to electricity production but could also supply all the industrial and domestic demand for heat at all temperature levels, their potential contribution to the energy economy could be much greater than we here calculated, perhaps a factor of 2 higher. This means that a little more than 0.1% of the land area could supply the bulk of mankind's need for heat and electricity. Note, however, that technology for the bulk collection, storage, and distribution of solar process heat does not yet exist. Nor would it be practical to have a purely solar energy system. The numbers quoted earlier merely illustrate the magnitude of the potential of solar energy.

5.2 Photovoltaic energy

The term *photovoltaic energy* is used to describe those processes in which solar radiation is converted directly to electricity without first being converted to heat or chemical forms. It is based on the property of certain solid materials, the semiconductors, to generate electrical current when illuminated by light of the appropriate wavelength.

The physical process underlying this phenomenon is the creation within the solid of free electrons upon the absorption of a light quantum. At the same time a "hole," a region of positive charge from which the the electron was freed, is also created. Unless special precautions are taken, such electrons and holes will recombine in a very short time and no useful current will be produced.

In photovoltaic cells the materials and their configurations are chosen so as to make the lifetime of the free electrons long enough to allow their collection and delivery to the outside circuit. The structural requirements on the atomic level of materials for photovoltaic applications and the way practical cells should be designed are well understood.

Every substance has a characteristic value for the minimum photon energy that will produce a free electron. Thus, for silicon the value is 1.1 eV (corresponding to a wavelength of light of 1.13 μm), and for gallium arsenide it is 1.4 eV (wavelength shorter than 0.89 μm). Because the solar spectrum extends from a wavelength of about 0.3 μm to over 3 μm, we can see immediately that not all the solar radiation can be used to generate free electrons.

The actual efficiency is much lower than would appear from the spectral considerations alone. The major factor that reduces the efficiency of conversion of light to electricity, after the free electrons have been produced, is recombination of those free electrons. Many mechanisms lead to the capture of the electrons before they have a chance to reach the outside circuit. Although recombination can occur in the bulk of the crystal, it is particularly severe at the surfaces and around impurities. Much of the art of producing efficient photovoltaic devices revolves around minimizing the recombination of electrons and holes.

An essential feature of the design of a photovoltaic cell is the creation within the material of a region possessing a strong electric field that will separate the electrons from the holes and lead to their collection. This region is generally known as the *junction,* and there are several ways of forming it. (See Figure 5.7.)

The electronic properties of a semiconductor can be modified greatly by incorporating minute (parts per million) amounts of foreign atoms into the crystal lattice, a process known as *doping.* By suitably choosing the dopant, one can create materials with an excess of electrons in their lattice by adding, for example, a small amount of phosphorus to a silicon crystal, or with a deficit of electrons in the lattice by adding, for example, boron to a silicon crystal. If different regions of the same crystal are doped in opposite senses, the contact between them will become a junction across which a potential will exist. The region of high electric field will draw any charges generated within it, or drifting into it, to one side or the other, and an electric current will flow in an outside circuit.

A junction can also be formed by bringing two different semiconductors, for example, cadmium sulfide and cadmium telluride, into contact. Other ways of forming junctions are also possible, each having specific characteristics that make it suitable for particular applications.

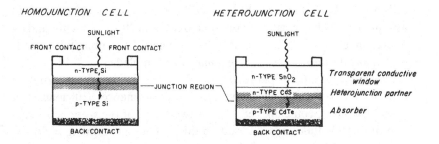

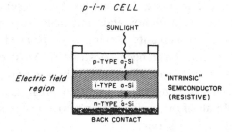

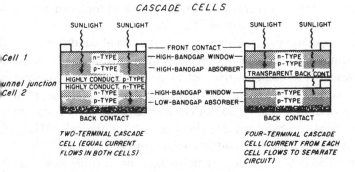

Figure 5.7. Schematics of various photovoltaic cells. (Source: Zweibel, 1986; copyright American Chemical Society.)

Single-crystal silicon. The "classical" material for the construction of photovoltaic cells is single-crystal silicon. Single crystals of high-purity silicon may be grown to fairly large sizes by carefully controlled slow cooling of a melt. Thin slices are then cut from the ingot of pure silicon and doped with the appropriate materials. Back and front contacts are attached, and the cell is then encapsulated for protection. Many such cells are assembled into modules, which are then incorporated into field assemblies. Each cell can deliver about 0.5 V and 20–30 mA/cm^2, and by interconnecting many of them in series and parallel combinations, modules of any desired electrical rating may be assembled.

Single-crystal silicon photovoltaic arrays are produced on a routine commercial basis, and they give reliable performance. Efficiencies of over 20% have been achieved. About two-thirds of the world's production of photovoltaic cells is based on this technology. Among the more familiar applications are satellite power supplies, remote communication systems, military installations, and a variety of isolated stations. Some of these have been in service for decades. More recently there has been an increased interest in photovoltaic electricity by public utilities, and numerous demonstration plants, ranging in size from a few kilowatts to hundreds of kilowatts, have been constructed. Most of them are based on the single-crystal-silicon technology. The largest plant is probably the 6-MW(e) facility constructed by Arco Solar Co. in Carrisa Plains, California. In all of these plants the photovoltaic power supplies give reliable and satisfactory service.

However, the cost of the electricity produced by this technology is too high for it to be widely used by the utilities. The main reason for this is the high cost of the cells themselves. Single-crystal silicon is an expensive material, and the manufacture of the cells, which requires highly skilled workers, must be carried out under "clean-room" conditions.

In an attempt to reduce further the costs of photovoltaic electricity, extensive research is being carried out to develop other photovoltaic materials and new technologies for their use. The general aim is to reduce the cost of materials by finding cheaper substances, by using less of each material, or by increasing the efficiency of the cells. These requirements are frequently in mutual conflict, and subtle tradeoffs have to be made. The cost of labor may be reduced by devising new, less demanding, technologies of cell fabrication. There is a bewildering variety of approaches to these problems, and some of them will be reviewed in the following paragraphs.

The cost of the cells themselves accounts for only a fraction of that of the completed assembly in the field. At the present time the support structures, tracking mechanism (if used), and power conditioning components may account for about half the total cost. It is clear, therefore, that further reduction of the cost of the photovoltaic cells only will have a progressively diminishing effect on the economics of the system.

Amorphous and polycrystalline silicon. To break away from the expensive single-crystal technology while staying with silicon as the base material, scientists have developed amorphous and polycrystalline forms of this element for photovoltaic use. These materials can be produced by techniques that are much simpler and cheaper than those for single-crystal silicon, and they can be applied in the form of thin films, thus requiring much smaller amounts. The efficiency of these materials for converting sunlight into electricity is lower than that of single-crystal silicon. It ranges between 10% and 12%, but it is hoped that this will be compensated for by lower production costs.

Amorphous silicon, in particular, has attracted much interest, and many successful devices using this material have been developed commercially. It is estimated that about a third of the total photovoltaic market in 1986 was based on amorphous silicon devices. Polycrystalline silicon is also easier to make and use in the form of thin sheets or ribbons, and shows much future promise.

Other directions being explored involve the search for more efficient photovoltaic materials and new techniques for fabricating them into devices. Among these materials gallium arsenide, cadmium telluride, and copper indium diselenide are the most promising. Gallium arsenide has yielded the most efficient photovoltaic cells yet produced, and they can be operated at high irradiation intensities and elevated temperatures.

Cascade cells. It was mentioned earlier that for each photovoltaic material there is a certain characteristic wavelength of light for which it is most efficient. A longer-wavelength light will not be absorbed. This circumstance can be used to increase the conversion of light to electricity by stacking thin cells, made from different materials, in such a way that light passing through the first will be partly absorbed by the second, and light emerging from the second will be absorbed by the third, and so on. Research today is concentrating on tandem cells, consisting of only two thin photovoltaic cells arranged in cascade. Much higher conversion efficiencies are obviously possible with such arrangements, but the design becomes increasingly complex. Many combinations of cells and their configurations are possible. A preferred type at present is a cascade consisting of a top cell of amorphous silicon (absorbing light with energy above 1.75 eV) and a bottom cell of copper indium diselenide (absorbing light above 1.0 eV). This should show an efficiency of 15%.

Here again the question is whether the extra efficiency is worth the extra cost and complication. The answer to this is not yet clear.

The common configuration of exposing the photovoltaic assemblies to solar radiation is in a form of flat plates mounted on a suitable frame. This can be fixed in a position that maximizes the total energy intercepted annually or mounted on a tracking device that keeps it facing the sun throughout the day. The gain in total energy yield of the latter configuration seems to outweigh the extra costs of the tracking device.

Concentrating cells. The total area of photovoltaic cells needed to generate a given power decreases as the intensity of the light increases. If the relation between illumination and power is linear, the area needed will be inversely proportional to the intensity of the light. Thus, for example, if the solar radiation can be concentrated 100 times before reaching the cells, only 1/100 of the area will be required, and the cost of this component correspondingly becomes much smaller. Actually, it has been found that the efficiency in some materials increases slightly with the intensity of the light, thus adding to the advantage. Experiments with single-crystal silicon, at solar intensities of up to about 500 times that of the sun, indicated conversion efficiencies of about 28%. With gallium arsenide single-crystal cells at 800 suns intensity, an efficiency of 26% was obtained. The concentrating devices usually considered for these applications are plastic Fresnel lenses mounted, together with the cells, on a tracking mechanism that keeps them aimed at the sun. One pays for this advantage by the expense of the concentrating and tracking devices and by the increased cost of cells suitable for high-intensity operation and for their cooling. A source of efficiency loss of concentrating systems, as compared with the usual flat-plate configuration, derives from the fact that the former can use only the direct solar radiation, whereas the latter can use the greater global radiation. Although several experimental photovoltaic power supplies based on this principle were constructed, its economic advantage has not yet been established.

Photoelectrochemical cells. It was pointed out in the course of the discussion of the mode of operation of photovoltaic devices that a junction between two materials that differ in their electronic properties is a crucial feature. Up to now we have described devices where the junction was formed within the solid. It is possible, however, to form a junction between a solid semiconductor and a solution. The cell formed in such a

way is known as a *photoelectrochemical cell,* and the interest in such
systems arose because of the possibility that the technologies involved
would prove to be simpler and cheaper than those of single-crystal solid-
state devices. The efficiency is expected to be somewhat lower, but this
may be compensated by the reduced costs. Variations of the basic design
have been constructed and tested. Efficiencies of about 8% have been
obtained readily, and in some special cases even 12% has been reported.
Engineering development and lifetime tests are still in progress, and the
future of this technology is still uncertain.

Space-based photovoltaic stations. Two disadvantages, common to all
photovoltaic systems, are the need for large areas of land, and the
intermittent nature of sunlight. Some years ago Glaser (1969) proposed
locating large photovoltaic assemblies in geostationary orbits in space.
The electrical energy generated would be converted to microwaves that
could be beamed down to earth, intercepted by large antennas, and
converted to electric current. Preliminary studies showed that the pro-
posal is feasible but unlikely to be implemented in this century.

5.3 Hydropower

In this and the following sections we discuss some of the indirect forms
of solar energy, that is, sources of energy that, although originating from
the absorption of solar radiation, have undergone some natural physical
or chemical changes.

The first source is *hydropower.* A part of the solar energy absorbed by
the oceans and the land masses causes some water to evaporate. The
rising vapor is entrained by the air currents (also driven by solar energy)
and carried to other parts of the globe, where some of it condenses,
precipitates on land, and ultimately runs back down to the ocean. Because this water starts its flow to the ocean above sea level, it possesses
some potential energy. This energy is exploited in hydroelectric schemes.

The utilization of the energy contained in falling or running water is
an ancient art. Throughout history and until the last century hydropower
was used solely as mechanical energy to drive mills and factories and to
pump water. Indeed, the pattern of the development of civilization and
human settlement followed in large measure the availability of hydropower. In its early forms the generation of mechanical power from
flowing water was a simple process and widely practiced.

With the introduction of electricity distribution networks, it became possible to uncouple the geographical linkage between the source of power and its ultimate user. This enabled the tapping of large additional resources of water power by the construction of huge hydroelectric projects, often in remote and inaccessible regions. We shall not dwell here on the technology of generating hydroelectric power, because it is a classical technology and well known.

The generation of hydroelectric power is a capital-intensive undertaking involving massive civil engineering projects that usually take a long time to complete. On the other hand, the fuel costs are zero, and the operation and maintenance costs are low. Hydroelectric plants are very flexible in their operation and eminently suitable for intermittent operation and peak-load supply. In this connection, we should distinguish between types of hydroelectric projects. Some depend entirely on the continuous flow of a river and have little or no storage capacity. Such systems are obviously suitable as base-load suppliers, because any water not used to generate electricity is "wasted" and cannot be stored for later use to satisfy peak demand. Other hydroelectric projects depend almost entirely on water stored in large reservoirs. In such systems any water saved can be used later, and it pays to conserve it for times of peak demand.

The unique characteristics of hydropower have also led to its use as a load management device to store off-peak surplus energy. Facilities constructed for such applications are known as *pumped-storage* systems. As the name implies, water is pumped up to a high reservoir at times of surplus energy and then dropped back through power generating turbines (often consisting of the same pumps used to fill the reservoir) when the demand increases.

Although the source of energy of hydroelectric stations is renewable, some of the projects to exploit it have a finite lifetime. The reason is that silting gradually reduces the storage capacity behind the dams (in projects that rely on reservoirs), thereby slowly decreasing their electricity production capacity.

The maximal global potential of this source of energy can be estimated from the knowledge of the precipitation on the land masses of the world and the average elevation at which the rain and snow fall. These numbers lead to a calculated absolute maximum energy of about 280 EJ fuel equivalent per year. This value is far above any practical possibility of exploitation, because the topographical situations suitable for hydroelectric projects are limited. It is estimated that the total world practical

Table 5.1. *Some major hydropower resources (in exajoules per year fuel equivalent)*

Country	Maximum theoretical capability	Exploitable capability	Actual generation
China	57.62	6.17	0.65
USSR	38.35	37.27	1.97
United States		6.5	3.18
Brazil	29.38	9.08	1.62
Indonesia	32.96	6.90	0.07
Canada	10.62	4.03	2.75
Japan	6.98		0.86
Norway	5.41	1.67	1.03
Mexico	4.86	1.68	0.23
India	3.86		0.47

Source: World Energy Conference (1986); U.S. Army Corps of Engineers (1981).

potential is about 90 EJ fuel equivalent per year (World Energy Conference 1986). Of this, only about 10% has actually been produced in 1984. The pattern of exploitation is very uneven and is much higher in the developed part of the world (65% of the potential) than in the developing nations (only 3% of the potential). Table 5.1 lists some of the major hydropower resources of the world.

The generation of hydroelectric power is not without adverse environmental impact. The huge construction project, the changes in the landscape, the flooding of land, and the changes in hydrological balance all have severe ecological implications. These will further limit the implementation of hydroelectric projects. The negative features are often partly compensated by improved amenities of recreation (boating and fishing in the artificial lakes) and by large benefits from flood control far downstream. In fact, the motivation for the construction of many of the large hydroelectric projects in the past was more as flood control devices than as energy sources.

5.4 Wind energy

The extraction of mechanical energy from winds is also an ancient art that has survived to this day. Windmills for pumping water can still be seen in many parts of the world. One reason for the survival of this technology is that it is particularly well adapted to the application and to the nature of the source. Small amounts of water are often needed in remote and unattended locations for irrigating small plots or as drinking water for cattle. The timing here is unimportant so that, whenever sufficient wind blows, it can be used and the water stored or applied directly. Furthermore, the technology is simple and within the ability of village craftsmen. The units are small, rated at not more than a few kilowatts, and many are operating in the world.

The situation is very different when we consider large power generating units meant for integration into an electric grid. The physical size of the units increases rapidly with the power, and the facility becomes an engineering challenge. Thus, the 2.5-MW(e) peak wind generator recently installed in California has a propeller that sweeps out a disk 91 m in diameter.

Because of the continued interest in wind power over many centuries and its rapid recent development, there is a feeling that the technologies are ripe for exploitation. Indeed, in many countries where the meteorological conditions are suitable, wind power presents an immediately available option for switching to a renewable energy resource.

Winds are driven by the uneven absorption of solar energy by the surface of the earth. Therefore wind is often classified as a form of solar energy that has been converted, by a huge (and low-efficiency) global thermal engine, to the kinetic energy of the moving air. Of the total solar energy intercepted by the earth, some 5.6 million EJ/yr, it is estimated (Gustavson 1979) that about 110,000 EJ are dissipated in the total wind system. Of this amount about 40,000 EJ are dissipated close to the surface of the earth. How much of this can be harvested without upsetting the earth's climate is not known. Obviously, too great a removal of energy from the wind would change the temperature and rainfall patterns of the globe and could not be tolerated. An accepted upper limit (which may be far too permissive) is about 10%. If we take this limit, we obtain a maximum possible wind energy extraction of about 4000 EJ/yr. This is still a very large number (compare it to the global energy use in 1985 of 320 EJ). The actual technical potential will likely be much smaller (taking into account economics and site availability), variously estimated at about 100 EJ.

The unpredictable nature of wind may pose serious problems in the integration of the energy generated in an electricity network. In this respect it is similar to some of the solar sources (photovoltaics, for in-stance). So long as the fraction of wind-generated electricity is small compared with the network's capacity, or more precisely of that part of the network that can respond quickly to demand changes, its incorpora-tion presents no special difficulties. But if the proportion of wind energy were ever to become an appreciable part of the network's capacity, severe problems in its management could be expected. These problems could be overcome by ensuring a sufficiency of quick-response standby generating capacity – an economic liability that would decrease the attractiveness of wind energy. Alternatively, wind-power plants could be coupled with some energy storage system, such as pumped storage. Such consider-ations, plus the constraints imposed by the availability of suitable geo-graphical and meteorological conditions, will probably limit severely the role that wind power may play on national or global energy scales.

5.5   Biomass

The term *biomass* is used in the energy context to describe all those sources that arise from recent photosynthetic processes. It covers sources such as agricultural and forest residues, industrial and municipal wastes, and crops specifically grown for conversion to fuels. Historically, biomass was the sole source of energy for mankind for most of its evolu-tion, to be replaced only in the last two centuries by the fossil fuels. At the present time biomass is again being examined as a substitute for these same fossil fuels that displaced it earlier.

Before dealing more specifically with the various forms of biomass, there are a few general remarks to make about the total potential of this source of energy. Because biomass arises from recent photosynthesis, its potential depends on the efficiency of this process and on the total area from which products useful to man may be harvested. The efficiency, in terms of the fraction of the solar energy converted to chemical energy, is generally low. Thus, of the close to 4 million EJ of solar energy reaching the surface of the earth each year, only about 3150 EJ are naturally converted to organic compounds by photosynthetic processes. Of this amount, about 38% are produced in the oceans, leaving some 2000 EJ equivalent of terrestrial organic matter synthesized per year (Bolin 1979). How much of this it is desirable or possible to harvest is a moot point. The most optimistic estimates place this fraction at not more than 40% (i.e., some 800 EJ equivalent). To achieve this value it would be necessary to cultivate practically the entire arable surface of the earth according to the most advanced practices.

A fraction of the 800 EJ of biomass must be set aside for the production of food and for materials directly dependent on plant growth, such as timber and paper, which in the next century could total about 380 EJ/yr. The efficiency of gathering and processing the remaining 420 EJ/yr of organic matter to usable fuels is also not perfect and probably averages about 50%, leading to a final annual potential of around 210 EJ equivalent. This value should be compared with the 320 EJ total world energy consumption in 1985. From these figures, we can see immediately that even for the present world population, a renewable energy system totally dependent on biomass is not realizable. At the same time recall that biomass will be the only source of naturally occurring reduced carbon when the fossil fuels run out. It will therefore be the base on which much of the chemical industry and liquid fuels production will have to depend in the future. Continued research and development effort in biomass as an energy and chemicals resource must therefore be maintained.

An important consideration in the production of all energy sources is the net energy balance. This is obtained by setting off the total, direct and indirect, investment in energy required to produce the resource to the stage at which it can be used, against the useful energy obtained from it. Energy resources differ greatly in their net energy balance, to the point that for some of them the balance may actually be negative. This is true for some of the forms of biomass, and we shall discuss this point where relevant.

In discussing biomass, one can draw a distinction between two major categories. The first includes all those sources that are by-products or wastes from other activities, for example, agricultural and forestry operations and urban organic wastes. The second category includes all those systems where plants are grown for the specific purpose of using them or products derived from them for the production of fuels or energy. A good example is the growing of sugarcane for the production of alcohol in Brazil. The potential and economics of the two categories are clearly different.

Most of the emphasis so far has been on the first category because of the more favorable conditions for its implementation. At first sight crop and logging residues may appear to be free, because they are by-products that have been paid for by the principal commodity. However, the collection and transportation of the residues to the point of use or to the processing plant presents a considerable cost item (estimated at $10 to $20 per tonne).

In any sustained utilization of crop and forestry residues, additional energy costs arise from the need to compensate for the loss of nutrients and soil conditioners otherwise provided by such residues. This occurs because the presence of a certain proportion of the residues, to be naturally recycled, is essential for the maintenance of soil fertility and the prevention of erosion. The fraction that may be safely removed from the fields or forests depends greatly on the nature of the crops, the topography, the agricultural practices followed, the climate, and other factors. It may vary from zero for crops such as sorghum and soybeans to over 70% for rice (the average for grains is about 25%). For the United States the average readily available crop residue is about 17%, and forest residues run at about 40%.

Plant residues and industrial and urban wastes may be used for energy production in many ways. Because such materials cannot stand, for economic reasons, long-distance transportation, they must be used in reasonably close proximity to their point of origin. The simplest method is by their direct combustion to raise process steam or generate electricity. An appreciable fraction of such residues is indeed used in this way. Thus the paper and pulp industry, the timber industry, and some food industries produce much of their heat requirements and some of their electricity needs by burning the various waste products available on their site or present nearby. This internal use cuts down on the general commercial availability of these sources.

Disposal of urban wastes by combustion, which simultaneously generates steam and electricity, is a well-known process that is finding increasing application. The global potential of this resource was estimated to be 28 EJ in 1975 and is expected to rise to 47 EJ by the end of the century.

The domestic use of firewood is ubiquitous in most parts of the world. It has been estimated that up to 50% of the global firewood harvested is used for heating and cooking, particularly in the less developed countries. (The proportion in these countries can reach up to 90%.) In much of the world the great reliance on firewood for domestic use has already led to extensive deforestation and created a fuel-wood crisis more severe than the oil crisis, because little oil was used. Lately, the proportion of domestic use of firewood has been increasing in industrialized countries. Naturally, the main areas in these countries where wood is used extensively as a source of domestic heat are those in the more rural and forested parts. Much of this wood is noncommercial and is gathered by the users themselves.

One of the simplest methods of processing wood to give a higher-grade fuel is its conversion to charcoal. The primitive method, widely practiced in antiquity and still in use today in many locations, is to burn wood with a limited access of air. The charcoal produced is a more concentrated and clean-burning fuel and may be shipped easily to domestic and industrial users.

A somewhat more sophisticated method, also practiced in the past, is to "dry-distill" wood by heating it in retorts. In this way valuable chemicals are obtained in addition to charcoal. One of these chemicals is methanol, and hence the common name "wood-alcohol" for this material.

More modern methods of thermal processing of wood involve its combustion with air or, better still, with oxygen, under carefully controlled conditions, to yield a gas containing mostly hydrogen and carbon monoxide (and nitrogen if air is used). This gas, which has a relatively low calorific value, can be used to fire boilers, run gas turbines for power or electricity generation, or drive internal combustion engines.

After proper adjustment of its composition the gas can also be used to synthesize methanol. The latter can be used as motor fuel, either alone or in admixture with gasoline.

Several biological processes exist for treating agricultural and forestry residues and urban and industrial wastes so as to recover valuable fuels from them. The best-known examples are the fermentation of carbohydrates to give ethanol and the production of methane by the anaerobic fermentation of farm residues.

Starch and cellulosic materials can be hydrolyzed by enzymatic processes or by treatment with acid to yield sugars in solution. These sugars may then be fermented to give ethanol. The yields are between 200 to 300 L of ethanol per ton of dry biomass. As far as energy balance goes, however, the net gain is not so great, and in certain cases may even be negative because of the considerable energy inputs involved in the collection, transportation, and processing of the material, and the indirect cost in energy of maintaining soil productivity.

Organic matter may be fermented under oxygen-free (anaerobic) conditions to yield a gas containing about 60 mol% methane, the rest being mainly carbon dioxide. The fermentation is carried out by a mixed population of fermentative, acetogenic, and methanogenic bacteria. Acting together these bacteria degrade carbohydrates to methane, carbon dioxide, and hydrogen. A typical yield is about 85 m^3 of methane per tonne of dry biomass, the equivalent of 3.3 GJ of energy. However, since an input of about 2.3 GJ is required for processing, the net energy gain is not more than 1 GJ, an efficiency of about 6% (based on 16 GJ/tonne as the energy content of the biomass).

A related process is the recovery of methane from landfill operation. Urban garbage is often disposed by burying it under thick layers of earth. Under these anaerobic conditions the organic matter in the garbage ferments slowly to give, among other products, methane. If the covering layers are impervious, the gas collects underground and may be tapped at a later date. Several such "natural" gas recovery projects have been implemented in recent years.

We may summarize this section by recalling that biomass from crop- and tree-growing residues can be harvested to an average proportion of 25% and 40%, respectively. This material can be burnt to yield process steam with an efficiency of 60% to 70% or used to generate electricity with an efficiency of about 25%. Biomass may be fermented to give ethanol with an efficiency of 12% or converted by anaerobic fermentation to methane with an efficiency of 5%.

The seemingly ready availability of valuable secondary energy forms such as electricity, liquid, and gaseous fuels from agricultural and forestry residues has generated interest in the possibility of growing crops and trees specifically for energy production. The considerations here differ in some respects from those employed in the discussion of residues. No longer can one regard the organic material as a by-product raw material of low cost. All the direct and indirect costs of growing the crops must now be charged to the energy produced. In addition, these energy plantations now compete directly with food production for resources and land.

The best-known example is probably the production of ethanol from sugarcane in Brazil (in 1983 some 5.5 million tonnes were produced). The bulk of this material (5.1 million tonnes) was used as motor transport fuel, displacing an equivalent amount of gasoline. Brazil is in a particularly favorable situation because its climate is suitable for growing sugarcane, which is among the crops yielding the highest fraction of biomass. The average yield of ethanol from sugarcane in Brazil is about 65 L per tonne of material. The energy balance is positive if the waste products (bagasse) are used to provide the heat requirements of the process. About 40% of the total energy value produced (alcohol plus residues) is used in the growing of the sugarcane and its processing to ethanol, leaving some 89 GJ/ha as net annual yield.

Crops containing a high proportion of starch may also be fermented easily to yield ethanol. In Brazil and in some African countries production of ethanol from cassava is contemplated. The yields from such a source are lower, and the net energy gain is about half that of sugarcane. Nevertheless, because this crop may be grown in areas not suitable for sugarcane, this route to ethanol is of some interest.

Another well-known example is the production in the United States of ethanol by the fermentation of surplus corn. The overall energy balance here is less favorable and in most cases is probably negative. In 1984 some 1590 million liters of ethanol for fuel were produced and used as 10% mixture in gasoline (gasohol).

Although ethanol may also be made from cellulosic materials (i.e., wood), it is likely that this process will be used mostly in connection with forestry and timber industry residues, and not with special energy plantations.

However, electricity generation by the direct combustion of wood may be based on energy plantations. According to this concept a specially planted and managed forest serves the need of a conventional wood-burning power station. It can be seen readily that a basic limitation on the capacity of such a power station is the distance from which it is still economical to transport the wood. Estimates place this at 80–100 km, which taken together with the average productivity leads to a maximum power output of 60 MW for the wood-burning electric power station. The possibility of planting special fast-growing species of trees to supply the needs of such power stations has been considered. Although there is little experience in this type of operation, it is probable that in the long run it will become possible to improve procedures and develop new varieties of plants that would lead to an appreciable increase in the power output.

Up to now we have confined our discussion to terrestrial sources. A large fraction of the global photosynthetic activity occurs in oceans and freshwater bodies. The harvesting of biomass from the aquatic environment for energy production purposes is, in general, not practicable. Exceptions are the use of kelp (a species of seaweed) for ethanol production, and the raising of algae in special ponds. The latter application is still in its infancy, and the main emphases of present developments are more as processes for obtaining chemicals (e.g., glycerol) than as an energy resource (Avron 1984).

5.6 Geothermal energy

All the energy sources discussed in this chapter are derived, directly or indirectly, from solar radiation. There are two additional potential energy sources not derived from solar energy: geothermal energy and the tides. We shall discuss these briefly in this section and the following one.

During the formation of the solar system, some 5 billion (10^9) years ago, small amounts of radioactive elements were incorporated into the primitive earth. These materials were probably present in the galactic dust clouds as debris from the terminal stages of the evolution of stars in the vicinity. Some of the radioactive isotopes, those with the longest half-lives, survive to this day. The most important ones for our present discussion are the isotopes of potassium ^{40}K (half-life of 1.26 billion years and an abundance of 0.118%), of ^{238}U (half-life of 4.47 billion years), and of thorium (14 billion (10^9) years). The energy release associated with their radioactive decay appears ultimately as heat in the surrounding rocks. This geothermal heat diffuses slowly to the surface of the earth, whence it radiates out to space. The average geothermal heat flow from the surface of the earth is 63 mW/m^2 or a global total of 1100 EJ/yr. As we shall see later, only a small fraction of this may be technologically available.

The heat flow through the bulk of the earth leads to a gradient of temperature between its interior and the surface, averaging in the crust to 30°C/km. It can be seen that with such a gradient rocks at the depth of the usual aquifers will be only a few tens of degrees hotter than the surface, and the water obtained from them will not be a very useful source of energy. In places where the crust is thinner and molten magma reaches closer to the surface, the temperature gradient can be much higher and useful temperatures may be reached even at depths of under 1000 m. This is the situation in regions of subduction of tectonic plates, recent mountain building, and rifting. Accordingly, we find that geothermal activity seems to concentrate in the circumpacific zone, the Mediterranean area, and the major rifts and midocean islands associated with them.

The energy of the hot rocks deep in the earth may only be extracted if they are in contact with water or brine solutions that may be pumped to the surface. An additional requirement is, therefore, that suitable aquifers be present in the geothermally active zones. This requirement greatly limits the number of potential sources of geothermal energy.

Depending on the temperature level, geothermal hot water may be used for heating hothouses, providing residential space heating, and generating electricity. Large-scale projects of geothermal space heating are in place in Iceland, France, Hungary, and other countries.

In some places the temperature of the aquifer may be so high that steam at pressure may be obtained directly from wells drilled into the strata and used to generate electricity in a more or less conventional way ("dry steam" type).

The water in high-temperature aquifers usually contains high concentrations of salts and minerals in solutions, is corrosive, and cannot be disposed of conveniently. In such situations it may be used to generate clean steam in special boilers or by direct flashing (i.e., allowing it to boil under somewhat reduced pressure). The remaining brine is usually reinjected into the aquifer. This procedure helps to maintain the balance of the aquifer and to avoid the environmental difficulties involved in the disposal of the brines.

In recent years a third method of tapping medium- and low-temperature geothermal sources has been introduced. The hot brine is used to boil an organic fluid, the vapors of which drive a special turbine (the "binary" type).

High-temperature geothermal sources have been exploited for electricity generation since the early days of this century, and a considerable generating capacity has been built up over time. Table 5.2 lists the countries possessing appreciable geothermal electricity generating capacity.

Assuming an average capacity factor of 80%, one can calculate that the world's geothermal electricity generation in 1985 amounted to 33 TWh(e), which is equivalent to a fossil-fuel consumption of about 0.3 EJ. Worldwide surveys of the practical potential of geothermal energy from such sources indicate a magnitude of about 100 EJ (EPRI 1985b).

Considerable additional potential could be tapped if new technologies for exploiting hot dry rocks could be developed. In many locations high-temperature rocks exist at depths within reach of modern drilling technology, but where no water is naturally present to act as a heat-transfer medium to bring the energy to the surface. It is possible to inject water into the borehole for this purpose, but then it is also necessary to ensure a large enough rock surface from which to extract the heat effectively.

Table 5.2. *Geothermal power plants operating in 1985 (countries with capacity greater than 10 MW(e))*

Country	Power rating [MW(e)]	Types
United States	2,022	Dry steam, flash, binary
Philippines	894	Flash
Mexico	645	Flash
Italy	519	Dry steam, flash
Japan	215	Dry steam, flash
New Zealand	167	Flash
El Salvador	95	Flash
Kenya	45	Flash
Iceland	39	Flash
Nicaragua	35	Flash
Indonesia	32	Dry steam, flash
Turkey	21	Flash
China	14	Flash, binary
Soviet Union	11	Flash
World[a]	4,763	

[a]Figure includes some countries with plants smaller than 10 MW(e).
Source: DiPippo (1985).

Research is in progress on technologies for fracturing the rock by explosives or other means so as to expose large surfaces and provide interconnected channels for fluid flow. It is clear that the productive life of such a well, and hence its economics, will depend greatly on the volume of rock that can be reached by the heat-removing water. This is essentially a "heat mining" operation in the sense that once the heat stored in the rock in the vicinity of the well has been extracted, its capacity will drop to low values because of the slow inflow of fresh geothermal energy. It is therefore not really a renewable source, but the total amount of geothermal heat stored in the upper 5 km of the earth's crust has been estimated to be approximately 40 million times the total reserves of oil and gas.

This problem is less severe where a reservoir of hot water is being exploited. Also in this case, overpumping may reduce the output of the well to lower values, but the initial amount of accessible stored energy is higher and its replenishment rate is faster.

5.7 Tidal energy

The gravitational pull of the moon and the sun, combined with the rotation of the earth, causes a tidal bulge in the ocean to travel continuously around the globe. The presence of continents and friction in the water dissipate some of the energy contained in the tides and slow the rotation of the earth. The energy dissipated (ultimately as heat) derives therefore from the angular momentum imparted to the solar system in its formation.

In some places on the globe the topography of the coastline is such that very high tides are created, and in these locations some of the energy contained in the tidal wave may be extracted as electricity. The technology is reminiscent of that of hydropower. Some of the water brought in by the high tide is impounded by a barrage and is then released back to the ocean at ebb tide through suitable turbines. The largest such project is at the La Rance River estuary in France, where nearly 240 MW(e) can be generated. Because there are few places on the globe where such schemes may be carried out, the potential is limited and estimated as about 1 EJ/yr.

Additional reading

Becker, M. (1987). *Solar Thermal Energy Utilization*, 3 volumes. Springer-Verlag, New York.

Bockris, J. O'M. (1980). *Energy Options*. Wiley, New York, p. 134.

Fahrenbach, A.L. and Bube, R.H. (1983). *Fundamentals of Solar Cells*. Academic Press, New York.

Green, M. A. (1982). *Solar Cells*. Prentice-Hall, Englewood Cliffs, NJ.

Hall, C. W. (1981). *Biomass as an Alternative Fuel*. Government Institutes, Rockville, Md.

Lynette, R. (1985). Wind Power Stations: 1984 Performance and Reliability. EPRI Report AP-4199. Palo Alto, CA.

OECD (1984). *Biomass for Energy*. OECD, Paris.

U.S. Army Corps of Engineers (1981). *National Hydroelectric Power Resources Study*, Vol. 12. Washington, D.C.

Wick, G. L. and Schmitt, Walter R., eds. (1981). *Harvesting Ocean Energy*. UNESCO Press, Paris.

6

Demand and substitution

6.1 Demand for energy

Before embarking on a discussion of energy demand and the substitution of one source by another, we mention two points. The final user actually requires a certain energy service, be it heat for the house, power to move machinery, or means of transportation, and he or she is not at all interested in the primary source that supplies this service. This gives rise to the possibility of substitution of primary sources, a subject to be discussed extensively later in this chapter. The other point relates to the usage of the term *demand*. At times when there are no external constraints, the demand equals use, and the two terms may be considered synonymous. If one ignores stockpiles, then the two terms are also equal to the supply. But when constraints appear, the demand for energy may not be equal to the actual use. Several things may happen when the demand cannot be met. At first, the users begin to compete for the supply, thereby raising its price. Then when this becomes too high, they may just have to do without the energy service or they may seek substitutes. Again, if we ignore changes in stocks, the actual use cannot be greater than the supply.

The problem of future demand estimates became pressing when such data were needed for the formulation of national and international energy policies. Very sophisticated procedures were developed to make energy forecasts, and we shall not attempt an exhaustive description here. For our present purpose we do not need precise estimates of energy demand, and the following brief review of some of the methods used in estimating it should provide sufficient background.

We shall limit ourselves to a time interval of about 50 years, because

115

the validity of the estimates decreases rapidly as the time span increases, and merges into prophecy. The simplest, and least accurate, method for estimating future energy demand is the historical approach. According to this method, the past behavior of the global energy system is analyzed, and the assumption is made that the system will continue to behave in the same way in the future. The historical record for 1870 to 1930 shows an average global growth rate of energy use of 2.3% per year (Putnam 1953). If this rate were to continue, the global primary energy use in 2030 would be 2.8 times the present and would reach 830 EJ/yr. The remarkable feature of this record is the approximate constancy of the growth factor in the face of major political upheavals, world wars, great social changes, and profound scientific and technological developments. It is likely that the very nature of the model, which aggregates the many diverse and sometimes conflicting factors contributing to energy use, somehow leads to internal compensation and a stabilization of the growth rate. The record for 1973 to 1983, when the importance of conserving energy became recognized, shows that the global growth rate of primary energy use dropped to 1.6% per year. At this growth rate the energy use in 2030 would be 2.04 times greater than it is today and would reach 613 EJ/yr.

Using this simple approach for the prediction of future demands leaves one uneasy, because the mechanisms governing the growth rate are not transparent, and one feels uncertain about assuming that the historical growth rate will continue. In fact, it is certain that growth cannot continue forever. For this reason other, slightly more detailed, approaches are sought, where some of the factors are isolated and their effects are studied and estimated.

The first step in disaggregating the various factors is to separate out the effect of population size. The total energy use is the product of the number of people and the per capita demand. Knowledge of these two factors and their time dependence will then give us a prediction of total energy demand. It turns out that the size of the population is a dominant factor in determining future energy demand. The analysis starts, therefore, with the estimation of the world's population at various times in the future.

Of all the factors, the world's population is probably the most amenable to extrapolation because it changes rather slowly, on the scale of human generation time, and because it is the least sensitive to the effects of the other parameters involved in the energy-use equation. But for all

Table 6.1. *Estimates of world population*

| | | Projection for year | | | | |
Source	Year	1975	2000	2030	2050	2075
Wilcox	1931		4.0		7.4	
Carr-Saunders	1936	2.9	3.6		6.0	
Wilcox	1940	3.0	3.9		7.2	
Putnam	1953	2.9	3.7		6.0	
Census	1975	3.98				
Chant	1981		6.08	7.98		
IAEA	1982b		6.09			
Keyfitz	1983		5.89	7.36[a]	8.20	8.45

[a] Value for 2025.

that, considerable deviations from predictions are found even for periods as short as 20 years. Table 6.1 shows some of the earlier estimates and the latest predictions.

In the following discussion we shall use the value of 8×10^9 for the world's population in 2030.

To continue the analysis, we now estimate the evolution of the energy demand per capita. Here again the simple historical approach could be applied as a first approximation and the global historical per capita energy use simply extrapolated into the future. Such an approach, however, would lump together countries with vastly differing per capita energy consumption and at various stages of economic development, and give grossly misleading results. Table 6.2 shows the great disparity in the energy consumption of different regions. There is also a great difference in the annual rate of increase of the per capita consumption. In North America the rate is around 0.3% per year, and in North Africa/Middle East it is 4% per year.

The data of Table 6.2 clearly show that it is first necessary to disaggregate the world into economic regions and then calculate for each of them the expected per capita energy demand as a function of time. The latter is usually based on the connection between economic development, measured in terms of the gross domestic product (GDP), and energy consumption. It is then necessary to make some assumptions regarding the probable course of economic development of each region,

Table 6.2. *Primary energy consumption per capita and population in various regions, 1985*

Region	Energy (GJ/yr)	Population (millions)
North America	311	263
Western Europe	128	406
Eastern Europe	185	419
Industrialized Pacific countries	141	140
Africa and Middle East	29	648
Asia	19	2536
Latin America	48	406

Source: International Atomic Energy Agency (1986).

and to further guess at the correlation between that course and energy consumption. In extrapolating such curves into the future, the implicit assumption is often made that the relationship between energy use and economic growth remains constant. However, when energy costs began to rise following the oil crises of the 1970s, increasing attention was directed to energy efficiency, conservation, and substitution, with the result that the traditional correlation between energy use and GDP was broken. It was found that economic growth could be maintained with a smaller rise in energy consumption and, for short periods, even with no rise. This was achieved in several ways, some requiring nothing more than careful management, others requiring changing processes and equipment. In other words, energy was traded for better organization and know-how.

In Figure 6.1 we show some projected regional correlations between primary energy consumption (E) and GNP taken from the work of Edmonds and Reilly (1985), where some of these factors have been taken into account. We note that, in general, the E/GNP ratio for the developing regions is expected to rise for the next 100 years or so while that for the developed countries is expected to decrease slightly. These trends can be understood in terms of the increasing industrialization of the developing nations and the increasing efficiency of energy use and conservation in the developed nations.

However, the complexity of making estimates of future energy con-

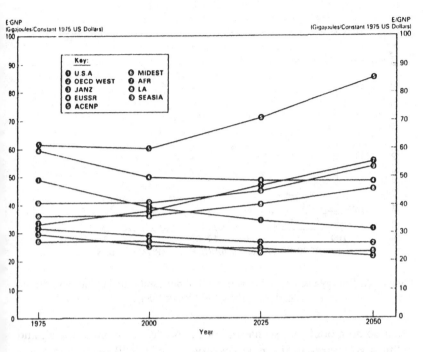

Figure 6.1. Energy consumption versus GNP for various regions of the world (GJ/constant 1975 $ U.S.) (Source: Edmonds and Reilly 1985, based on research at Oak Ridge Associated Universities.)

sumption is brought out clearly in plots such as Figure 6.2, where are collected the data for the energy consumption per capita versus GDP per capita for most nations. The general correlation between affluence (as measured by the GDP per capita) and energy demand is clearly visible, but there are significant deviations. We notice in Figure 6.2 that countries of approximately similar affluence may differ greatly in their per capita energy consumption. Conversely, we see that countries with similar per capita energy consumption may differ greatly in their per capita gross national income.

The reasons for this are many and are not simply related to differences in energy-use efficiency, as might seem at first glance. The detailed analysis is complex and beyond the scope of this book (see the reading list at the end of the chapter).

The historical and other correlation approaches may be of some use for short-range predictions, but their precision decreases rapidly with the time range. Any discussion of future energy demands therefore

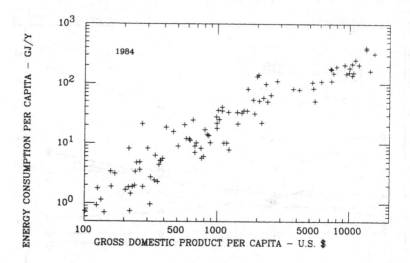

Figure 6.2. Energy use per capita versus GDP per capita for 111 nations. (Based on data of the World Bank 1984.)

needs some model, or "scenario," of the world's economic, social, and political structures in the next century and of the likely technological developments. In Figure 6.3 we show a selection of the results of some studies of possible future energy demands for the next 50 years.

The wide scatter of the points for the first third of the next century is not the result of errors in extrapolations or different techniques of calculation. Rather, it results from the different models of the future world used in these studies. An infinite number of such scenarios may be concocted, and, correspondingly, any desired hypothetical future demand pattern may be derived by choosing the appropriate model.

There may be two reasons for engaging in the construction of such scenarios. The first may be the belief that some of these models are more likely to approximate to future reality and, therefore, be useful for quantitative predictions. The second reason for constructing such models is to help us understand the importance of the factors that determine energy demand. Such an understanding is required for the formulation of international energy policies, should developments make such a thing possible. It is in the spirit of the latter reason that the discussion of scenarios is presented in this and the following chapters.

The need to postulate scenarios for future energy demand analysis introduces a strong subjective element. The political and social beliefs of

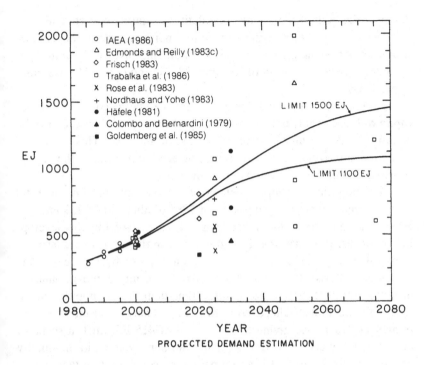

Figure 6.3. Estimates of future global primary energy demand.

the analysts and their hopes and visions for the future are bound to be reflected in the inputs to the models used in the calculations. It is probably this fact that causes most of the difficulties in the debates on energy, and the polemics and acrimony that ensue.

For the broad discussion of future energy demand, the possible ways of meeting it, and their environmental consequences, a detailed or precise scenario is not necessary. However, because a certain scenario is the basis for much of the discussion that follows, a declaration of my assumptions regarding the future may not be out of place. I believe that a world with a large disparity in the quality of life between regions and between countries cannot be maintained. Although it is certainly possible to achieve a high and satisfactory standard of living with less energy consumption per capita than at present in North America (over 300 GJ/yr), it is impossible to approach anywhere near a satisfactory way of life on 20 GJ/yr. It is highly unethical to demand, and unrealistic to expect, that global energy use be frozen at today's level. This is tantamount to condemning the bulk of the world's population to eternal poverty and even to a possible *reduction* in the per capita energy use in the developing countries. As Table 6.2 shows, 66% of humanity lives with annual per capita energy consumption of less than 30 GJ, and only 5.5% use more than 300 GJ.

We therefore make the assumption that the energy consumption per capita will rise steeply in the less developed countries, which are also the countries with the highest rate of population increase. This effect will dominate and overshadow any freeze, or even reduction, in the energy consumption of the developed part of the world.

Some numerical examples will illustrate the options. Of the total world primary energy consumption in 1985 of about 319 EJ, some 215 EJ were used by the industrialized countries and 104 EJ by the developing countries. If we assume that no increase in the per capita consumption will take place in the future, the energy use by the developing countries will rise to about 200 EJ as the result of the population increase alone. If we further assume that there be no population increase in the industrialized countries, their energy consumption will remain at 215 EJ/yr, giving a world total of 415 EJ. Such a scenario means that any economic development and improvement in the quality of life can only occur if it is accompanied by compensating technological developments in energy use, and probably by societal changes. It is a minimalistic scenario, probably unrealistic and unacceptable to the developing nations.

A fairer scenario assumes that the developing nations will strive to advance as rapidly as possible and, as a consequence, increase their energy consumption greatly. One can make various assumptions as to the ultimate per capita energy consumption to be reached by these nations. We see no reason why it should not be similar to that enjoyed by Western European nations in 1985, about 128 GJ. In this case the annual energy consumption of the developing nations will reach 880 EJ. If we still restrict the growth of the energy use per capita of the industrialized nations and allow only for the projected increase of their populations, we obtain an estimate of annual energy use of 270 EJ. The limiting global energy use will thus be 1150 EJ.

If we repeat this calculation but now assume that the developing countries will reach the 1985 level of per capita energy consumption of Eastern Europe (185 GJ), we obtain a global annual demand of about 1500 EJ. The discussion in this and the following chapters is based on an ultimate world annual energy consumption in the range 1100 EJ (low) to 1500 EJ (high). With these simple assumptions, it is possible to obtain an appreciation of the magnitude of the problems to be faced if an attempt is made to attain the levels of energy availability postulated in these scenarios.

To proceed, we make the basic assumption that the future growth curve of energy demand does not rise indefinitely but flattens out at some future date. In other words, we assume that the annual energy-use rate is constrained in the future even if the resources are not. This is consistent with the often-made assumption that the world's population will reach an approximate steady state in the future, as will the per capita consumption. Accordingly, the curve describing the future annual demand for energy will have an S shape, showing an initial exponential growth and a final asymptotic approach to some steady state, L EJ/yr. Many mathematical forms show this behavior, and we have chosen the logistic function to represent it because of its simplicity and its later usefulness in describing energy substitutions. This use of the logistic function differs from that frequently employed to describe the fraction of a constrained resource exploited. For our purpose the logistic equation can be written as

$$f(t) = \frac{\exp(\alpha t - \beta)}{1 - \exp(\alpha t - \beta)}$$

where $f(t)$ is the production rate at time t expressed as a fraction of the annual limit L, and α and β are constants.

The limit L represents the steady-state level of the total primary energy demand one assumes for the far future. The constant α determines the rate at which the steady state is approached, and the constant β determines the time scale. The constant α is chosen to reproduce the assumed growth rate of primary energy consumption in the next 25 years. The constant β is determined from the annual energy consumption rate at the starting point of the calculation (usually the present). These parameters permit us to calculate possible demand curves beyond the range of the estimated values. Calculated demand curves for L in the range 1100 to 1500 EJ are also shown in Figure 6.3. The procedure for carrying out such calculations (and those in Chapter 7) is outlined in Appendix B.

The curves in Figure 6.3 apply to the total energy demand and cannot, therefore, be used in comparisons with the individual resources that

Table 6.3. *Percentage shares of various types of fuel, 1985*

Region	Solids	Liquids	Gases	Hydro	Nuclear	Geo
North America	24.34	38.98	23.87	7.52	5.17	0.12
Western Europe	19.92	45.76	15.97	8.09	10.20	0.06
Eastern Europe	34.35	29.92	30.02	3.22	2.45	0.05
Industrialized Pacific countries	21.00	54.25	12.20	5.00	7.42	0.13
Asia	65.96	23.85	4.67	4.49	0.89	0.14
Latin America	24.54	44.42	16.47	13.99	0.42	0.17
Africa and Middle East	43.83	41.94	11.19	2.76	0.27	0.02
World total	33.45	37.00	19.19	6.05	4.22	0.09

Source: International Atomic Energy Agency (1986).

contribute to it. To carry the analysis further, we have to consider the fractional contribution of each energy resource to the total supply. Table 6.3 shows the percentage share of primary energy sources in the total consumption in 1985. We see that 37% of the energy consumed derives from oil, and 56% from oil and gas together. In the industrial countries the proportions of these fuels are even higher, reaching over 62% for oil and gas.

By way of an exercise we can now attempt to compare the cumulative demand for specific primary energy sources with the magnitudes of the resources. To do this, we assume for the moment that the pattern of energy use is frozen at the 1985 proportions shown in Table 6.3. We can then apply these proportions to the global demand and calculate the cumulative consumption of each kind of fuel. This can then be compared with the magnitude of the relevant resource. These comparisons are shown in Figure 6.4.

Figure 6.4 shows that under these assumptions conventional oil will not last much beyond the year 2020, and the unconventional oil will add only about 20 years to this lifetime. The conventional gas resources will similarly be unable to meet the demand beyond about 2040, but if the unconventional sources are included the gas may last to the end of the next century. These conclusions are independent, within a few years, of whether 1100 or 1500 EJ are chosen for the limiting annual energy consumption.

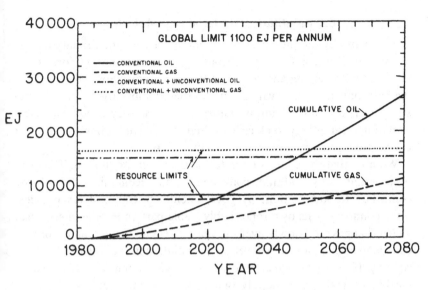

Figure 6.4. A comparison of cumulative demand and resources of oil and gas.

Although Figure 6.4 is based on the unrealistic assumption of the constancy of the present fuel-use fraction, we feel that the general conclusions are not affected, and that one should therefore expect a massive substitution of conventional oil by other energy sources in the first half of the next century. It is probable that long before these critical dates an increasing fraction of the demand for liquid fuels will be satisfied by a mix of materials produced from unconventional oil sources and from oil shale and coal. The specific mix will depend on the economic competitiveness of the products and will vary greatly from region to region. Globally, the proportion of conventional oil use will go down while that of coal will go up, and new sources such as unconventional oil and oil shales will also contribute.

We have seen that such substitutions may not always be technologically feasible, and some may have undesirable social and environmental consequences. The substitution process may be limited by insufficient knowledge and may require great investments of financial, material, and human resources. The ability to meet the rising demand for energy will therefore be limited more by the difficulty of fuel substitution than by the availability of other primary sources. In the following sections we shall discuss these problems in greater detail.

6.2 Substitution: the technological challenges

The demand by the final user is seldom directly for the primary energy source; usually it is for some secondary or tertiary energy form (i.e., after some processing and one or more conversion steps). The processing and conversions can vary greatly in their complexity, from the fairly straightforward purification of natural gas to the very complex conversion of nuclear energy to electricity. Figure 6.5 vividly shows the flow of energy from primary sources to final consumers.

We note that the choices open to the final user are limited. They will be determined by tradition, convenience, and economics, and may be controlled to some extent by governmental pricing and tax policies. Each country has its own remarkably stable pattern of final energy use, which changes only slowly over the years. It is safe to say that any drastic change in such a pattern will reflect a revolution in the way of life of the country. If such a change is imposed by outside circumstances, for example by the inability to supply one of the energy forms to which the society had become used, a crisis situation will arise.

The utilities engaged in the production and distribution of final energy sources have, in general, a greater choice as to which primary sources or combination of sources to use. For example, the electric utilities have a relatively broad choice, because all primary energy sources can be used to generate electricity. For other secondary energy producers, the choice is more limited. Thus, the refineries that produce petroleum products can only choose between grades of crude oil and the processes by which to treat them. All utilities will try to maximize their income by the proper mix of the primary energy sources available to them. The optimal mix will change with time according to economic conditions and technological developments.

In the discussion so far, we made the implicit assumption that the technologies for energy production and use exist, and the choice between them is based primarily on economic reasoning. This is proper for a description of the present and near-future situations. But unfortunately this happy situation is not universal and is not applicable for discussion of future trends. In the chapters dealing with the particular primary energy sources, we have alluded where possible to the open questions that remain before the resource can be utilized and replace oil. We shall review these problems in the following pages and amplify where necessary.

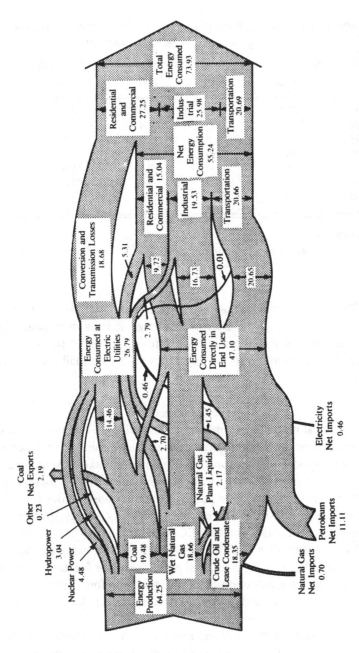

Figure 6.5. Energy flow diagram for the United States. (EJ = 10^{15} BTU. Source: *Annual Energy Review* 1986.)

Table 6.4. *Percentage share of primary energy sources in the generation of electricity (OECD countries)*

Year	Solid	Oil	Gas	Nuclear	Hydro
1960	47.2	7.2	13.2	0.2	32.0
1970	39.5	19.7	14.4	2.1	24.3
1973	35.2	25.0	13.5	4.4	21.9
1975	36.7	22.0	11.2	7.5	22.6
1980	21.0	17.0	12.0	11.0	21.0
1985	41.5	8.0	9.9	20.6	20.0
1990[a]	39	10	7	25	17

[a] Estimates.
Sources: World Energy Conference (1983), International Energy Agency (1986).

We start by recalling that we are interested in the substitution of oil and other fossil fuels by alternative energy sources, particularly by renewable or inexhaustible ones. The different energy sectors vary greatly in their adaptability to source substitution, and we shall review them in turn.

6.2.1 Electrical

The classical methods of generating electricity are by burning oil, coal, or gas to generate steam to drive turbines, and from falling water. Geothermal and wind energies are also beginning to make some minor contributions. An electric utility will choose a mix of such sources as are available to it, primarily on purely economic considerations. Table 6.4 shows the contribution of the various primary energy sources to the generation of electricity in the OECD countries.

We note that oil is already being substituted for in the electric sector by coal and by nuclear power, and is not the major fuel today. The main interest in fuel substitution in this sector, therefore, arises more from environmental considerations (acid rain and carbon dioxide effects) than from fears of resource depletion.

Of the more recent sources, nuclear power, although economically competitive with fossil fuels in most countries, involves additional considerations that slow up its introduction into many power networks. These problems were discussed extensively in Chapters 3 and 4 and will not be repeated here. Recall, however, that the opposition to nuclear power is based partly on fears relating to safety, radioactive waste disposal, and environmental considerations. These issues require further research and development efforts for their resolution, and from that point of view the technology may still be regarded as immature. It was further pointed out that the present, once-through, light-water technology does not really offer a very long term solution in terms of resource availability. At the moment it seems that only a breeder technology or fusion, if developed to the point of general acceptability, will provide an inexhaustible source of power. Here again much further development is necessary. If these conditions are fulfilled, then nuclear energy will become a true substitute for all the fossil fuels in the electric power generating industry.

Of the other possible sources for generating electricity, the renewables, mainly solar energy in its various forms, direct and indirect, are still all in the development stage. They are still economically inferior to the classical sources and to nuclear energy, and require considerable development effort, but taken together they can provide an appreciable fraction of mankind's needs in the future on a permanent basis. To summarize, although there are a number of substitutes for fossil fuels in the electric sector, most still require further development.

6.2.2 *Transportation*

Transportation is a major oil-consuming sector: 41% of all oil used in OECD countries in 1980 was in this sector. It is also the most difficult energy-consuming sector from the point of view of the replacement of oil by other sources.

The development of the transportation field was dominated by the plentiful availability and convenience of liquid fuels. A whole new way of life has evolved in this century, based on the great personal mobility provided by motor transport, which will be very difficult to change radically. Much of the commerce of nations also depends extensively on internal combustion and jet engines, as well as on oil-burning steam turbines. Extensive fuel processing and distribution systems have developed, and vast road networks have been constructed. Because of all this, there will likely be a strong pressure to maintain liquid fuels as a base for transportation energy. The onus of substitution thus falls on the fuel industry, on the refiners, and on the processors of primary fossil fuels. They have to find new ways of making liquid fuels from materials that differ from those hitherto used by them. The difficulty of doing this will increase with time. The shift from conventional to unconventional crude oil probably involves only minor changes in the processes, and the required technologies either exist already or can be developed by the industry in a reasonably short time. They can probably be implemented with investments that are within the capability of the industry.

A temporary expedient that can be used to eke out the supply of motor fuel is to use the heavier components of natural gas (LPG – mainly propane and butane) directly in internal combustion engines. The modifications needed to engines running on LPG are not great, and some experience with such conversions already exists (Canada, Australia, New Zealand, Thailand, and Japan).

A similar source of alternative motor fuel, although less convenient, is compressed natural gas (CNG – mainly methane). It is used in Italy and New Zealand on a commercial basis. Both of these fuels are of particular interest to countries possessing large natural gas resources or distribution systems.

But the major sources of alternative crude oil with properties similar to conventional oil will undoubtedly be coal, tar sands, and oil shale. These materials can be processed to yield ultimately liquid fuels suitable for transportation and other uses. Processes for doing this may be divided into two classes: those depending on the prior conversion of coal (or oil shale) to a gaseous mixture from which the liquid fuels are then produced by a further reaction, and those that produce a liquid product directly.

Many coal gasification processes differing in the details of construction and operation have been developed. Basically they all produce a mixture of carbon monoxide and hydrogen. This mixture is a low-BTU fuel and can be used as such. But the main value of this mixture of gases, and the one that also gives it its name, *synthesis gas,* is its use as a starting material for a whole range of synthetic organic products. By varying the reaction conditions and the catalysts used, we can produce methane, methanol, higher hydrocarbons, and a large variety of oxygen-containing organic compounds. The earliest use of one such process, the Fischer–Tropsch process, was developed in Germany just before World War II and was used extensively during the war to produce liquid fuels. It is still used on a large scale in South Africa under the name of the *Sassol process.* Several other processes have been developed since then and have reached pilot-plant or demonstration scale. It is likely that when justified by economic conditions, similar or improved versions will be deployed in suitable locations, probably near large coal mines. One can regard, therefore, such processes as alternatives to the present liquid fuel sources when they become scarce. Another route to coal-derived liquid fuel is through the direct treatment of the mineral with suitable solvents, under elevated temperature and pressure and in the presence of catalysts. Several processes have been developed and some have reached the pilot-plant stage, but no commercial production by any of them has yet been established.

The processes discussed so far yield familiar liquid fuels, so they will not involve the final user in changes of equipment or habits. It is also possible to produce from coal, oil shale, and natural gas, liquids that may be used as motor fuels but which have characteristics different from the conventional oil derivatives. One such material is methanol. In Chapter 2 we described the production of methanol from natural gas and its use as motor fuel. The same substance may be produced from coal by the indirect route (i.e., via synthesis gas). In view of the much larger reserves of coal, it is this substance that may be the source of methanol and derived fuels in the future. If methanol does become a future fuel distributed to the end-user, some minor changes in the internal combustion engines will become necessary, but basically no great difference in the traditional way of life will result.

From the discussion so far, it is apparent that there are fossil substitutes for oil in the transport sector and that the technologies for their use exist or are in an advanced state of development. Although this situation seems to answer the short-term problem of oil depletion, it does not provide any relief from longer-term problems arising from carbon dioxide emission. Replacing one fossil resource by another actually aggravates this problem because the extra processing steps entail additional carbon dioxide emission.

If we broaden our discussion to include the search for substitutes for all fossil fuels, with the aim of minimizing the environmental effects, we come up against a host of new and less investigated problems.

The liquid fuels considered so far, the hydrocarbons and the alcohols, are all organic, carbon-containing materials. It is for this reason that their combustion forms carbon dioxide. There is, however, one additional source of organic liquid fuels – biomass. Biomass are materials derived from atmospheric carbon dioxide by photosynthesis driven by solar energy (see Section 5.5). For this reason, their combustion does not add to the net carbon dioxide load of the atmosphere. It follows that liquid fuels derived from biomass are entirely benign, as far as carbon dioxide emissions go.

A variety of liquid fuels may be made from biomass (see Chapter 5), including ethanol from the fermentation of sugar or starch and methanol from the chemical processing of wood. The best-known example of large-scale application of such processes for fuel use is that of Brazil, where extensive ethanol production from sugarcane was initiated in the 1980s.

It is only when such options have been exhausted or become unavailable that major problems will face the transportation sector. Several choices are still open in such a case. One choice is to fall back on inorganic sources of carbon to produce liquid organic fuels. The carbon could be obtained from atmospheric carbon dioxide, from carbonate rock, from flue gases, and as a by-product of fermentation. The hydrogen could come from water by electrolysis or some other process, the source of energy being nuclear or renewable.

Another choice is to abandon carbon-based fuels altogether and switch to an entirely different chemistry. Hydrogen in various forms is frequently mentioned in this connection. In such a case not only will the engines be different but the whole distribution network serving the transportation sector will have to be changed.

An extensive electrically driven public transportation system, coupled with the development of a practical electric car, is another possible solution (provided the bulk of the electricity to drive the system is generated from nonfossil sources). Although no particular technological problems exist with the electrification of public transport, this is not true with the electric car. No electric car that would find favor with the general user has been developed yet although special-purpose vehicles have been in use for some time. The major problems are limited range and speed and the excessive weight of the storage batteries. Sea and air transport are not amenable to such solutions. Research is continuing on these topics, and one can hope that before the problem will become severe acceptable solutions will be forthcoming. An extensive electrification program of the transportation and other sectors will reduce the demand for oil and may postpone the need to introduce the electric car.

Nuclear power has been demonstrated as a feasible mode of propulsion for oceangoing vessels. However, apart from military and special-purpose craft (ice breakers) this technology is not generally acceptable (see Section 4.4).

Table 6.5. *Share of energy carriers*
for industrial uses in OECD
countries, 1980 (%)

Oil and	42
oil products	
Gas	22
Coal	18
Electricity	17
Others	1

Source: OECD (1980).

Table 6.6. *Proportion of delivered energy used for*
various industrial tasks in OECD countries (%)

Process steam	25
Direct process heat (less than 600°C)	18
Space and water heating	12
Metal melting and heating	5
Electrolytic chemicals	2
High-temperature direct heat 600°C	10
Iron and steel	20
Stationary motive power	8

Source: OECD/IEA (1982).

6.2.3 Industrial

The industrial sector uses energy in secondary and tertiary forms. The share of energy carriers for industrial end use is shown in Table 6.5. We note again the predominance of oil in this sector. Considering that two-thirds of the electricity supplied is also generated from fossil fuels, we see that over 90% of the energy used by industry is of fossil origin. It is instructive to examine the end uses of the energy supplied to industry. Table 6.6 shows a breakdown of the average use of energy for industrial functions in OECD countries.

There are also nonenergy uses of fossil fuels. Thus, the petrochemical and fertilizer industries rely heavily on petroleum products and natural gas as their basic raw materials, and the metallurgical industry uses coal as a reducing agent. In some of these applications substitution by other materials is possible even with present knowledge. Thus, hydrogen for ammonia production need not be made by the steam reforming of hydrocarbons or from coal, but it can be prepared electrolytically. Similarly, the reduction of iron ore can be performed with hydrogen rather than with coke. In the petrochemical industry there is, however, no substitute for oil.

In many of the industrial applications, particularly those involving heat, substitution among the various fossil-energy sources is possible without too great an investment or changes in the processes. We note from the first three entries in Table 6.6 that about 55% of the total energy needs of industry fall into this category. The mix of fossil sources to satisfy these needs will be determined by considerations of cost and convenience. The substitution of oil by coal does present some environmental problems, which may be overcome at additional costs. These problems relate both to the transport of coal and to its combustion in an environmentally acceptable manner. Research on these topics is in progress, and the future of coal's share in the industrial sector depends in great measure on the successful outcome of these efforts. Most of the requirements of the remaining categories in Table 6.6 can also be satisfied by a combination of fossil fuels, but perhaps not so readily.

It is when we consider the possibility of replacing all the fossil fuels by renewable resources that we reach the limits of present technological experience. The magnitude of the problem is evident from the data in Table 6.5.

An obvious solution is to increase the degree of electricity use in industry, provided that it is generated mostly from nonfossil sources. The technology for the direct electrical heating of industrial processes, and of electrical raising of process steam, is available and is used to some extent. But in general such an approach is economically less advantageous and is also less conserving in total energy. The main reason for this is the great thermodynamic penalty associated with the generation of electricity from a heat source. If electricity is produced from coal or gas, say, and then converted to heat by the final user, that user will obtain only about 30% of the heat of combustion of the fuels for his or her process. If the same fuels were used directly in the user's process, he or she would obtain 2–3 times as much useful heat (see Section 1.4). This penalty may be acceptable, if the source of heat is nonfossil, as the price of avoiding carbon dioxide emissions.

There are several ways to ameliorate the apparent wastefulness of using electricity to generate heat. We first mention the method of cogeneration. This is the arrangement by which the final user generates his or her own electricity and uses the waste heat from the turbines in the process. Special turbines are available for such applications, which can provide heat at levels suitable for many processes (at some sacrifice in the efficiency of generating electricity). The overall efficiency of fuel use, counting both the electrical energy and the process heat, is much higher than if each were generated separately.

The second method is the use of heat pumps. Heat pumps are devices that may be likened to refrigerators, or air conditioners, run in reverse. They use mechanical energy to extract heat from the environment or other low-temperature source and deliver it to the user. The efficiency, called the *coefficient of performance,* of this process is expressed as the ratio of the heat pumped to the electrical energy invested. It depends strongly on the difference between the temperature of the source of the heat and that of the process to which it is delivered: the greater the difference the lower the efficiency. Because of this characteristic, heat pumps are useful mainly in low-temperature applications, such as space and water heating. Values of the coefficient of performance ranging from 2 to 6 have been obtained in industrial applications. If we consider that the efficiency of generating electricity is, on the average, about 33%, and if this is used to drive a heat pump with a coefficient of performance above 3, the net result is actually a better utilization of the heat source than if it were applied directly to the process.

The direct use of nuclear power and other renewable resources in industry is more complicated. It is, in principle, possible to consider nuclear energy as a substitute for fossil fuel in industry. Reactors can be used to generate high-pressure and high-temperature steam and there-fore could be a source of process heat and all lower-temperature heat loads of industry without much modification. However, because of safety and other reasons, which were discussed in Chapter 4, it is unac-ceptable at present to locate nuclear reactors close to densely settled regions. These are just the places where the industries needing the heat are located. Siting the reactors at a distance from the users of its heat is not practical because of the inefficiency of transporting sensible heat over long distances. It is for these reasons that nuclear energy has, so far, not been used directly in industry.

There are two ways of solving this difficulty. The first is by developing inherently safe reactors that, hopefully, could be sited without opposition even in inhabited areas, close to the potential industrial user. The other approach is to develop systems for the indirect transport of nuclear or solar heat with higher efficiency than can be obtained by its prior conversion to electricity. The extensive research carried out in Germany on the thermochemical pipeline concept is an attempt at solving the latter problem. According to this concept, an endothermic chemical reaction is carried out at the high temperature of the reactor output. The products of the reaction can then be transported to the point of use, where the reaction is reversed, regenerating the starting materials and releasing the heat absorbed in the initial forward step. The cycle is closed by returning the reaction products back to the reactor. High-temperature gas-cooled reactors have been considered for this application.

Several chemical systems could be used in such a scheme. The most studied so far is steam-reforming of methane. In this process, steam and methane are reacted at high temperatures (around 900°C) in the presence of a suitable catalyst. The reaction proceeds with the absorption of a large amount of heat (220 kJ/mol methane) to give a mixture of hydrogen and carbon monoxide. This reaction is well known in the petrochemical industry and is a standard way of producing hydrogen. The mixture of gases at ambient temperature may be piped to any distance and could be stored on the way. At the user's end the mixture of hydrogen and carbon monoxide is made to react, again in the presence of a catalyst, to give methane and steam, the original starting materials, with the evolution of the heat absorbed initially. The net effect of this cycle is to transport, and store if necessary, in the form of chemical energy the energy initially supplied as heat. The concept has been proved in a 10-MW pilot plant (using electric heating) at the KFA, Jülich, Germany (Harth and Boltendahl 1981). The coupling to an operating high-temperature gas-cooled reactor has yet to be demonstrated. The successful demonstration of this concept can open the way to a considerable contribution of nuclear energy to the industrial process heat demand. In this way the total contribution of nuclear energy to the economy can be increased over and above its contribution through the electric sector alone.

The same concept may also find application in the field of solar energy (see Section 5.1). The main handicaps facing solar energy arise from its uneven distribution and its intermittent nature. Before solar energy can be considered a serious substitute for fossil fuels in the industrial sector, the problems of its transportation and storage will have to be solved. These are reminiscent of the problem discussed in the previous paragraphs in connection with the possible use of nuclear reactors as a source of industrial heat. In both cases, although for entirely different reasons, the source of energy cannot be very close to the heavy user. It is therefore possible that a solution found suitable for the nuclear case would also be useful in the solar application. In the latter case such processes have an additional attraction because they also provide an answer to the storage problems. It is for these reasons that the thermochemical pipeline concept is being studied in several laboratories for its solar energy applications.

In the absence of such a general solution, which could permit the setting up of a national or even international solar energy distribution network for industrial, commercial, and domestic heat, the field of application of this plentiful renewable resource will remain limited to the electrical sector and, in a few cases, to some special other uses.

Of the remaining renewable resources, water and wind were used in the past as the main sources of mechanical energy in industry. The siting of mills and the villages and towns that grew around them were determined largely by the availability of rivers and streams that could be harnessed for power production. It is unlikely that there will be a return to such systems, and therefore direct water and wind power cannot be considered as a substitute source of energy for industry.

6.2.4 Domestic and commercial

The domestic and commercial sector accounts for 32 to 42% of the primary energy used in the world. In the industrialized nations this energy is supplied by three carriers: electricity, petroleum products, and gas. In the developing countries up to 50% of the energy may be supplied by biomass. Of the petroleum products, the main fuels are the medium distillates (e.g., kerosene, light fuel oils). A small proportion of heavy fuel oil is used for steam raising. The use of coal in this sector has been all but phased out in favor of other energy vectors because of convenience and environmental constraints. A breakdown of the main categories of end use is shown in Table 6.7.

A breakdown of the energy carriers serving the domestic/commercial sector in some industrial countries is shown in Table 6.8. Examination of Table 6.8 shows that the fossil fuels represent the bulk of the energy carriers in the industrialized countries. Replacing fossil fuels by electric-

Table 6.7. *Major end-uses of energy in Western Europe and the United States in the domestic/commercial sector, 1980 (%)*

	Western Europe	United States
Space heating	80	66
Water heating	10	17
Cooking	5	3
Electrical appliances	5	14

Table 6.8. *Carriers of delivered energy to the domestic/commercial sectors in some industrial countries, 1982 (%)*

	European community	United States	Eastern Europe	USSR
Oil	41	28	20	19
Solids	9	–	48	23
Gas	32	43	10	18
Electricity	15	28	9	11
Other	3	1	13[a]	29[a]

[a] Mainly steam and hot water.

ity in space heating, cooking, and hot-water supply is only a matter of economics and convenience. There is no technical difficulty in achieving an all-electric house, and in some locations they are quite common. There are still some advantages of convenience and economics in using gas for heating and cooking, but the switch to electricity is not difficult. The considerations discussed in connection with industrial process heat are also relevant to the domestic and commercial sectors. Here again, if electric power is to be produced from nonfossil sources, the increased domestic and commercial electrification is equivalent, in effect, to substitution by renewable or nuclear energy.

The discussion of ways of compensating for the inherent thermodynamic inefficiency of converting heat to electricity and back again to heat, which was presented earlier in connection with the industrial sector, is valid for the domestic/commercial sector as well. A combined heat and power generating plant supplying electricity and district heat will have a higher overall efficiency than if these services were provided by separate plants. Because the combined plant must be located close to the users of the heat, it will probably not be accepted if the energy source is nuclear, even though such a system has been operating in Sweden for many years without trouble.

If the concept of the thermochemical heat pipe is ever implemented in industry, it could without difficulty also supply district heat to the domestic and commercial sectors.

The heat pump is eminently suitable to the low-temperature heat loads of private and public buildings. Heat pumps, usually combined with an air conditioning function in one unit, are now available commercially. The technology of heat pumps is well developed, particularly in the larger sizes of pump, and its use is increasing slowly. The main obstacles to the greater use of heat pumps in the domestic sector are initial costs and some lack of convenience (noise, for example). If the design is improved and large-scale use is encouraged so that a mass market could develop, a major step toward the drastic reduction in the consumption of fossil fuels in the domestic sector could result.

There is also considerable scope for the direct use of renewable energy resources in the domestic/commercial sector. By the intelligent use of proper architectural design, it is possible to reduce the required heating and cooling loads of buildings. In addition to this passive approach, it is possible to use solar energy to provide the hot-water demand and some of the space heating load. The first application is extensively used in some countries, and there is no reason why it could not be extended to others, particularly as simple systems, with no moving parts, are often sufficient. The use of solar energy for space heating is hampered by the lack of satisfactory diurnal and seasonal heat storage systems. When such systems become available and competitive with electric heating (with or without heat pumps), it will in principle be possible to transfer a major fraction of the total domestic and commercial energy requirement to a renewable source.

We shall not discuss in this section the question of the relative merits of producing the electricity locally (for example, by small wind or photovoltaic devices) as compared with centrally generated current from renewable or nuclear sources.

Most of the considerations presented here are valid only for industrialized countries. In the developing countries, the main energy carrier to the domestic sector is biomass. The problem there is how to ensure an ample supply of this fuel and use it efficiently and conveniently. Small-scale and local electricity generation from renewable sources can help greatly in improving the quality of life. These problems will be discussed further in Chapter 7.

6.3 Substitution process

We have shown earlier in this chapter that a major replacement of oil by other energy sources is expected in the first half of the next century. Initially, this will occur by the exploitation of unconventional oil and oil shales and by the liquefication of coal. Later, it may become imperative to substitute renewable or inexhaustible sources for all fossil fuels. All these substitutions require some research and development effort. In some cases this effort is modest, as in the case of the almost fully developed processes for coal liquefication, but for most of the renewable and inexhaustible sources, it is an enormous undertaking. Some of the technologies for these processes are not fully developed yet, and others are still in a very rudimentary, or even nonexistent, state. The future of the energy field is thus dominated by many technological changes, some very profound, that may affect our way of life, the balance of world power, and the environment.

At this point we should ask whether this present lack of knowledge and experience should concern us today. After all, we are dealing with a situation that will develop in the future. The answer to this question obviously depends on the time required to develop and deploy a major new technology. Can we therefore say something quantitative about the time needed for the substitution of one resource for another and for the introduction of new technologies? This question is important because it defines the time scale by which we should judge whether the depletion of a resource is an urgent matter and whether a crisis is likely to develop.

In the course of technological development the replacement of one process by another or of one raw material by alternatives is a common and expected event. For energy sources the historical record also shows such replacements in the past (see Figure 6.7). Considerable attention has been recently devoted to the dynamics of the substitution process of one technology for another (Fisher and Pry 1970; Marchetti and Nakicenovic 1979). Analysis of historical records revealed a remarkable

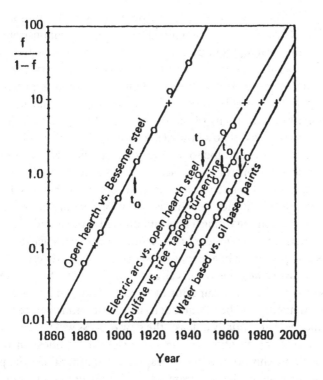

Figure 6.6. Examples of some technological substitutions. (Source: Fisher and
Pry 1970.)

regularity and similarity between diverse technologies. Figure 6.6 shows
examples from various fields.

It was found that over a good part of the life-span of a given technology, its rise and decay can be represented by the logistic function already
mentioned in Section 6.1. This function can be rewritten as

$$\ln\left(\frac{f}{1-f}\right) = (\alpha t - \beta),$$

where f is the fraction of the market captured by a given technology, t is
some time scale, and α and β are coefficients to be determined from the
historical record. When plotted on a semilogarithmic graph, such expres-
sions are straight lines. We note that the pattern is quite general, and
that a period of 30–50 years is required for the substitution or introduction of a new technology. There are good reasons for this behavior.
First, there is the natural process of developing an idea to the point of its

practical applicability – that is, to the point when its technological and economical feasibility on a commercial scale can be demonstrated. This section of the development path is in itself necessarily long; it involves extensive laboratory studies, then pilot plant tests, and finally the operation of one or more demonstration plants. Each step is progressively more expensive, roughly by an order of magnitude, and more time-consuming. Further, at any of these stages, problems that set the development process one or two steps backward may be discovered.

Experience shows that the total duration from the inception of an idea to its practical demonstration is measured in decades. At that point the new process has not yet penetrated the market for which it was designed. The process of substitution itself is also slow for several reasons. There is the natural caution regarding a new and untried process and a reluctance to be the first. Because there is little accumulated experience, an element of risk is also present. Then in many cases some other process or technology that achieves the same results already exists. Unless the new technology is so overwhelmingly superior as to justify the scrapping of a good plant, the tendency will be to wait out its natural life-time before replacing it. This means that, initially at least, the new process will be implemented mainly in new construction and in expansion of facilities. In times of economic boom this process may be rapid, but at other times the process may take the pace of the replacement of obsolescent equipment. Thus, this phase of a market penetration to a significant degree, usually taken as 50%, also takes several decades. So altogether the introduction of a major new technology can take 30 to 50 years.

The historical experience supports this analysis. We can take a couple of examples from the energy field itself. The example of nuclear power is particularly significant. It is a subject that received enormous support from governments, East and West, and rode on the tails of high-priority military programs. It can therefore be taken as an extreme example of an accelerated development process. And yet what is the record? Fission was discovered in 1938, the idea of a self-sustaining chain reaction formulated soon after (1939), its feasibility demonstrated on December 2, 1942, and the first power reactor operated in 1953. This is very rapid progress indeed, made possible only by the military implication of the subject. Yet today, almost 50 years after the discovery of fission, one cannot say that fission power has penetrated the world's energy economy to a major extent. It is true that in some countries, notably France, fission power accounts for over 12% of the primary energy needs. But,

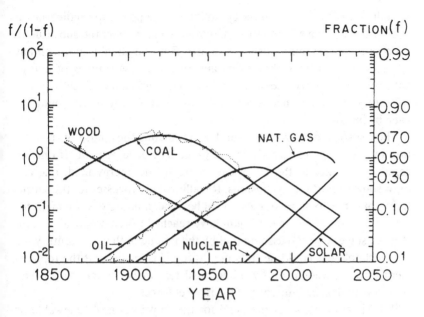

Figure 6.7. Historical substitution of primary energy sources. (Source: Marchetti and Nakicenovic 1979.)

in general, the subject is in retreat under a feeling that the technology has yet not been perfected and that more development is necessary (see Chapter 4).

Another example from the area of new technologies is fusion. In the mid-fifties the idea of fusion-derived electric power was touted as the most promising, and the views were so optimistic that it was even suggested that the further development of fission could be dispensed with, especially because fusion appeared to be much "cleaner" and it does not require uranium. Now 30 years and several billions of dollars later, even the scientific feasibility of the process has yet to be demonstrated. Thus one has to accept the long lead times for the development and deployment of major alternative technologies.

From plots such as Figure 6.6 it follows that a few historical points (theoretically only two) are sufficient to describe and predict the behavior over a long time. Figure 6.7, taken from Marchetti and Nakicenovic (1979), illustrates this behavior for the world's primary energy sources. The insensitivity of these plots to wide changes in economic and political events is remarkable. However, the model cannot predict the time of introduction of a new technology, so it has little power to predict the far future. After all, a still unborn technology cannot affect the substitution process until very many decades hence. But one should approach any discussion of future technologies and possible new sources of energy with proper humility. Less than a hundred years ago radioactivity was still unknown, and what is a major source of energy today could not even be imagined.

The logistic model may be useful for predicting the course of substitution by technologies already in the process of development. If one accepts the validity of the historical record to the energy field, one can draw some important conclusions. Recalling that processes for the substitution of other fossil resources for oil have been under development for some time, and that some of them have reached the commercial demonstration stage, one can assume that by the middle of the next century the new technologies could be fully deployed. This means that the replacement of conventional oil by other fossil resources is likely to proceed smoothly under the influence of the market forces.

But when we consider the need for the substitution of renewable or inexhaustible energy resources for fossil fuels, the situation is not so sanguine. The technologies needed here are almost nonexistent, and it is very difficult to guess the course of their development. Certainly the full period deduced from historical experience, and then some, is likely to be needed. Whether or not this is serious depends on the outcome of present research into the climatic effects of carbon dioxide emissions. Should there be no objection to the continuation of burning fossil fuels, then the existing resources of unconventional oil, coal, and oil shale probably provide sufficient time for the normal evolution of new substitution technologies. But if it should turn out that it is imperative that the use of fossil energy sources be curtailed, then the rate of the development of the new technologies will become the controlling factor in determining the severity of the carbon dioxide effect.

There is a qualitative difference between the substitutions we are now discussing and those that have occurred in the past in various technologies and are still proceeding there. The usual motivation for introducing new technologies has been purely economic – that is, because the new technology or a new resource is more efficient and more competitive than existing ones. The process of substitution then takes its course under market forces without major political or national policy implications.

In the energy field, however, we are dealing not only with substitutions resulting from the general advance of technology but also with substitutions forced by external constraints and with time schedules imposed by resource exhaustion or environmental factors. If the rate of introduction of the substitute energy source is limited by the development process, then a crisis situation may develop.

These features of the substitution process in the energy field dictate a different approach. No longer can one rely on substitution driven entirely by present market forces, for they will come into operation much too late to allow for the development of the required technologies. And in some cases, such as those forced by environmental considerations, there may be no market forces at all to drive the developments. Therefore these major substitution processes will have to result from central policy decisions on national and international levels. Chapter 7 deals with this subject more fully.

For energy we have only one historical example of substitution under pressure of source exhaustion. The substitution of coal for wood, in England at least, occurred in a crisis situation. The resource used until then was running out, and alternatives were sought. But coal, which had been known and used on a small scale, was already available to substitute for wood, and the conversion to this source of energy was rapid (within 50 years).

The substitution of oil for coal in this century was not driven by a shortage crisis but by its greater convenience and economic advantage. Coal was, and still is, plentiful.

Additional reading

Grenon, M. (1981). *The Nuclear Apple and the Solar Orange*. Pergamon Press, New York.

Taylor, R.H. (1983). *Alternative Energy Sources for the Generation of Electricity*. Adam Hilger, Bristol.

Winter, C.-J. and Nitsch, J. (1986). *Wasserstoff als Energieträger*. Springer-Verlag, New York.

World Energy Conference (1983). *Oil Substitution: World Outlook to 2020*. W.E.C. Conservation Commission. Graham & Trotman and Oxford University Press.

7

The missing resource

In the preceding chapters we saw that sources of energy, alternative to fossil, exist in amounts that could be sufficient for mankind's needs for all the foreseeable future. Thus, fission based on uranium and breeder technology could sustain an energy use seven times the present for about 100,000 years. Fusion based on the D-T reaction could do so for a million years, and the D-D reaction has sufficient raw material for 4 billion (10^9) years. In addition, there is a steady flux of solar energy, which in its various forms can also supply a demand 300 times greater than the present one for as long as the sun shines. These resources can therefore be regarded as practically inexhaustible. So far no other permanent energy sources have been identified, and therefore the future transition from the present, essentially fossil-based system, to one based on nuclear and solar sources seems inevitable. In Table 7.1 we collected the various energy sources together with their potential.

However, this optimistic conclusion must immediately be severely tempered by the realization of the difficulties inherent in the substitution of the present sources by new and inexhaustible ones (see Chapter 6). The simple reason is that we lack the knowledge and technologies to make a universal use of these permanent sources.

The energy problem is therefore not a potential shortage of sources but a shortage of knowledge of how to harness these sources. In fact, the usual presentation of the energy problem, in terms of the magnitude of the resources only, may have helped to distract attention from this main issue. This attitude probably results from insufficient appreciation of the dynamics of the substitution process and from the belief that when society will need new information and new technology, it will somehow always become available. It is possible that the very success of science

146

Table 7.1. *Some energy resources*

Type of resource		Reference
A. Nonrenewable, finite resource (EJ)		
Conventional oil	8,000	Table 2.1
Unconventional oil	7,000	Table 2.2
Gas	7,400	Table 2.5
Unconventional gas	> 9,000	Section 2.4
Coal	65,000	Table 2.6
Oil shale	19,000	Table 2.3
B. Renewable (annual capacity; EJ/yr)		
Solar energy	> 93,000	Section 5.1
Wind	4,000	Section 5.4
Hydro	90	Section 5.3
Geothermal	100	Section 5.6
Tides	1	Section 5.7
Biomass	210	Section 5.5
C. Nonrenewable, practically unlimited capacity (EJ)		
Fission	$> 200 \times 10^6$	Section 4.3
Fusion	$> 2000 \times 10^6$	Section 3.8

and technology in the last few decades has generated the confidence that the continuation of the process is guaranteed, and that therefore there is no need to worry about it explicitly.

In the absence of a global policy, it must be assumed that each nation will develop its own energy strategy. The mix of energy sources and the research and development efforts related to their utilization will be optimized to each country's specific needs and conditions. By and large the present pattern of energy use of each nation is most likely to persist, and the net effect of such a "business as usual" situation will be that the world will continue to remain largely dependent on fossil fuels for the near and intermediate future.

However, well before an appreciable depletion of conventional oil occurs, each nation will discover that it has to start producing its own substitutes or to arrange for supplies from sources of unconventional oil. Globally, there will be a shift from conventional oil (and gas) to the less conventional and synthetic sources (heavy oil, tar sands, oil shale, and coal). This massive shift away from the present conventional liquid fossil

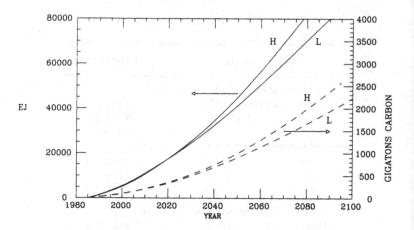

Figure 7.1. Calculated cumulative fossil-fuel use and carbon dioxide emission, with no definite substitution policy, 1985–2100.

fuels is expected in the second quarter of the next century (see Chapter 6 and Figure 6.4), but about 90% of the energy used would still continue to be derived from fossil fuels. The implication of these assumptions on cumulative fossil-fuel use is shown in Figure 7.1.

Figure 7.1 was drawn under the same assumption of a logistic behavior of future energy consumption as was made in Figure 6.4. The cumulative fossil-fuel-consumption curves of Figure 7.1 were calculated under the assumption that the fossil-fuel fraction remains constant at 1985 values. We note from Table 7.1 and Figure 7.1 that on the basis of these assumptions, the cumulative use of fossil fuels would approach the combined total fossil resources (115,000) early in the twenty-second century. This would seem to be a comfortable conclusion, because it appears to indicate that there is still plenty of time available for the development and deployment of the various technologies needed for the transition to permanent non-fossil-energy sources. However, closer examination shows that such a sanguine attitude is not justified. In drawing Figure 7.1, we assumed that conventional oil and gas could be readily substituted by other fuels derived from fossil sources and that there would be a smooth transition from one fuel source to another. These particular substitutions require the massive introduction of new technologies on a much shorter time scale than that relating to resource exhaustion. With the help of Figures 6.4 and 7.1 (and conditional on the assumptions made in plotting them), one can set limits on the time available for the deployment of the new technologies. From Figure 6.4 one can deduce that by the year 2060 the cumulative demand for oil and gas together will just exceed the total of the conventional and unconventional resources of these fuels (about 31,500 EJ). At the same time the total cumulative use of fossil fuels (Figure 7.1) will reach about 50,000 EJ. We can take this magnitude of fossil-fuel use as the limit of the resources available without the large-scale introduction of synthetic fuels (gas and liquid) and oil shale.

Although some of the technologies for the transitions to these new sources are already available, others are yet to be developed. If started in good time, they would probably not constrain the smooth shift from one fossil source to another in response to the needs. To remind the reader of some of the problems that have to be solved in this connection, we shall mention the need to develop processes for the production and treatment of the unconventional oils, including probably in situ processing, the development of a large-scale oil shale industry, and the development of an extensive synthetic fuels industry (see Chapter 6). All these will have to be deployed within the first half of the next century and will heavily strain human and material resources.

No sooner will this phase of the evolution to a coal- and shale-based energy economy be completed, than time will begin pressing for the transition to the more permanent nonfossil system. According to this picture, therefore, there will be no respite from the pressure of development and introduction of new energy technologies throughout the next century. In view of this it seems worth considering whether such a sequential transition from one basic technology to another, with a relatively short lifetime of the intermediate ones, would be the most sensible policy. Might it not be, perhaps, more economical in human and economic resources to press directly to the system based on the "permanent" energy sources? The total time and effort devoted to the research, development, and deployment of processes in the energy field could then be much reduced simply because of the fewer intermediate technologies that would have to be mastered. According to this view, the research and development efforts in the energy field should be directed toward the early introduction of renewable and inexhaustible resources as substitutes for the present fossil-based systems.

There is, however, another factor of a different kind, which also imposes a severe constraint on the time available for the development of substitute technologies. It is the possibly serious environmental conse-

Table 7.2. *Carbon dioxide emissions from various fuels (Tg carbon/EJ)*

Fuel	Preparation	Combustion	Total
Conventional oil		19.7	19.7
Shale oil	27.9	19.7	47.4
Gas		13.8	13.8
Coal		23.9	23.9
Synthetic oil (from coal)	18.9	19.7	38.6
Synthetic gas (from coal)	28.9	13.8	40.7

Source: Marland (1982).

quence of the extensive combustion of fossil fuels. Each fossil fuel generates a certain proportion of carbon dioxide upon combustion. In Table 7.2 these values are shown, expressed as the amount of carbon (in the form of carbon dioxide) emitted per unit of heat energy (in exajoules) produced.

In the processing of coal or oil shale to give unconventional oil or synthetic oil or gas, a considerable amount of energy is used. This leads to an accelerated use of the fossil resources and to the extra release of carbon dioxide to the atmosphere. The effect, therefore, of using synthetic fuels and shale oil is to increase greatly the amount of carbon dioxide released per unit energy supplied to the final user. This can be seen clearly in the data of Table 7.2.

If, in the future, technologies that do not depend on fossil energy (for example, those using solar or nuclear heat) will be developed for processing coal and oil shale, then synthetic gas or oil will not contribute more carbon dioxide to the atmosphere per unit energy delivered than the present conventional fuels. This may be one way of slowing down the buildup of carbon dioxide in the environment. Remember that the use of biomass does not increase the atmospheric carbon dioxide because any carbon dioxide emitted on combustion or processing of biomass was borrowed from the atmosphere a relatively short time before. Unfortunately the potential of this source of energy is limited (see Table 7.1).

The mix of fuels used at any time determines the rate at which carbon dioxide is emitted per unit energy delivered. At the present time, when most of the energy is derived from oil, coal, and gas, the release of carbon dioxide averages 19.7 million tons/EJ of combined fossil heat. As the shift away from conventional to nonconventional fossil fuels proceeds, the average amount of carbon dioxide released per unit energy delivered to the final user will rise and ultimately reach that of the synthetic fuels.

The calculation of the increase in atmospheric carbon dioxide involves two stages, both of which are somewhat complicated. It is first necessary to calculate the total amount of carbon dioxide emitted using some models for the total fossil-energy consumption and for the fuel mix as a function of time. It is then necessary to estimate the fraction of the carbon dioxide that is retained in the atmosphere. It is only the latter amount that is important in relation to the climatic effects.

Much work has been devoted to this problem in recent years, and many models have been devised to estimate the total energy use and fuel mix at various times in the next century and to derive from them the carbon dioxide release rate. The results of these calculations vary over a wide range, reflecting the different assumptions and different models used. A detailed discussion of this work is out of place here (Trabalka 1985).

The carbon dioxide emitted into the environment partitions between the atmosphere and the oceans, and there is still some uncertainty about the fraction of the gas retained by the former and its dependence on the total amount and the temperature. We have used the value of 0.6 for the atmospheric fraction in the calculations reported in this book.

It is apparent that the prediction of the atmospheric carbon dioxide levels in the far future is a very uncertain matter at the present state of knowledge. But for the next few decades one can assume that the pattern of energy use will not change drastically, and therefore somewhat better estimates can be made. Most recent studies converge on the prediction of the doubling of the atmospheric carbon dioxide around the middle of the next century. However, we can make some useful observations without getting involved in the fine details of such calculations. From the data of Tables 7.1 and 7.2, we can calculate readily that if all the liquid and gas resources were burnt, some 520 gigatons of carbon (in the form of carbon dioxide) would be emitted, of which some 310 gigatons would be retained by the atmosphere. This will raise the level of the carbon dioxide to 1.4 times its present value (335 ppm or 710 gigatons). We can also calculate that if all the recoverable coal resources were also used, the level of carbon dioxide would rise to about 3 times the present value. These numbers, which do not depend on any model or scenario, will be found useful later in the discussion of various policy options.

We described earlier an example of a "business as usual" scenario on the basis of which we plotted Figure 7.1 to represent the corresponding cumulative fossil-fuel use. Adding now a model of the fuel mix for the same period (which we have taken from Häfele 1981 and Edmonds and Reilly 1983c), we can compute the expected cumulative carbon dioxide emission. These results are also plotted in Figure 7.1, where we note that when 1185 gigatons of carbon have been emitted into the atmosphere the concentration of carbon dioxide in it would double (710 gigatons retained). We can read off the plot that this would happen around the middle of the next century, in agreement with many other calculations. We also note that this "doubling date" is not very sensitive to the assumed asymptotic level of the annual energy consumption (1100 or 1500 EJ).

Although there is a consensus about the inevitability of the increase in atmospheric carbon dioxide in the future, there is still a great uncertainty about the exact consequences of such an increase (see discussion of the greenhouse effect in Chapter 2). Nor is it likely that there is a critical level of carbon dioxide below which nothing will happen and above which dire effects will occur. However, it is generally accepted that one should try not to exceed a carbon dioxide level twice the present one, at least until such time as the whole subject of climatic effects of atmospheric trace gases is better understood. We shall use this value in the following discussion, with the caveat that it should be taken as a convenient reference level only.

From the simple computations presented earlier, it follows that the evolutionary path leading from the conventional liquid and gaseous fossil fuels to the coal- and shale-based synthetic fuels is inconsistent with the constraint of limiting the rise of the atmospheric carbon dioxide concentration to below its doubling level. The total effective resource level of the fossil fuels that may be exploited is therefore much smaller, probably around 50,000 EJ. By coincidence this limit is close to the cumulative amount of fossil fuels that will be used by the middle of the next century (See Figure 7.1).

In Chapter 6 (Figure 6.4) we saw that on a "business as usual" scenario (i.e., in the absence of a definite global energy policy) the oil and gas resources will be nearing exhaustion by the middle of the next century, so some action to develop substitutes in good time will become imperative. From the discussion on the preceding pages, it should be apparent that there are two directions that future energy planning can take, each with its own research and development program and its own timetable. One direction is to develop replacements for the liquid and gaseous fossil fuels based on coal and oil shale as raw materials. The other direction is to proceed directly to the renewable or inexhaustible sources. Which path to follow depends to some extent on the weight given to the environmental and climatic effects of carbon dioxide emissions. For if one attaches little significance to the deleterious consequences of continued fossil-fuel use or assigns a small likelihood to their occurrence, then the first path of energy evolution through the nonconventional fossil sources and coal- and oil-shale-based systems would seem to be acceptable. We repeat that the penalty for such a policy, in addition to the risk of causing an irreversible damage to the environment, is the need for a long and sustained effort to develop and deploy the technologies of coal- and shale-based energy systems in addition to those based on renewables and nuclear sources that will ultimately replace them.

If, on the other hand, a grave view is taken of the threat to the environment and the earth's climate from the continued burning of fossil fuels, then a direct transition to systems based on non-fossil-energy sources would be indicated. In this case a more vigorous research on, and development of, the renewable and inexhaustible sources is essential.

Put in another way, the choice is between going directly for the ultimate energy system based on renewable and inexhaustible sources or developing and deploying in addition an intermediate fossil system based on coal and oil shale. These choices face us irrespective of the carbon dioxide problem. The latter only reinforces the advantage of the direct transition to the nonfossil system. In either case the new technologies must be ready to take over the load toward the middle of the next century.

In Chapter 6 it was shown that, in principle, the transition to an essentially non-fossil-energy system is possible, although the information needed to do so is not fully available at present. This conclusion would be encouraging, if it were certain that the necessary knowledge could be gained in time. In the next few pages we examine this question in greater detail and attempt to define the various constraints more precisely.

In the following discussion it will be helpful to consider separately the industrial and the developing countries. We could then set different values for the growth of energy use and possible rates of substitution for each group of nations.

In 1985 the average per capita energy use in the industrialized countries was 201 GJ, and that for the developing countries was 28 GJ. In the spirit of the discussion in Chapter 6, we shall assume no long-range growth of the per capita energy consumption in the industrialized countries. For the developing countries we shall assume a more rapid growth of energy use to a limit of either 128 GJ per capita (about the level of Western Europe at present) or 185 GJ per capita (the present level of Eastern Europe). These values, taken together with the population projections for 2075 (Keyfitz 1983), probably close to the future stable levels, lead to an estimate of the ultimate total annual energy needs of the industrialized countries of 270 EJ, and for the presently developing countries either 880 or 1270 EJ (in rounded numbers). It is from these considerations that we use in our calculations the values of 1100 and 1500 EJ/yr as defining the range of the limiting total global annual energy consumption.

Because of the much higher future energy consumption by the developing nations that we have assumed, we expect that the extent and rate of fuel substitution in these countries, and not events in the industrialized nations, will be the determining factor in the rate of rise of the carbon dioxide concentration.

To obtain an appreciation of the magnitudes involved, we show in the following figures the calculated cumulative fossil-fuels use and the corresponding carbon dioxide emissions for assumed values of the relevant parameters. We assume a logistic behavior of the total energy use and of the rate of introduction of the new nonfossil sources. For the industrialized countries we again assumed essentially no growth in the per capita energy consumption, with the total energy-use growth following the rate of increase of the population. This population growth was taken to be about 0.5% per year over the next 25 years. For the developing countries we have assumed a much greater rate of growth both of the per capita energy use and of the population, leading to an overall initial annual energy growth rate of some 5% over the next 25 years.

Because it is obvious that fuel substitution, with its very demanding technological effort, would be easier in the industrialized countries than in the developing ones, we have examined what would happen if the

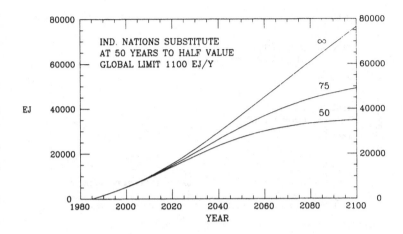

Figure 7.2. Calculated, cumulative, global use of fossil fuels, 1985–2100, for various rates of substitution in the developing countries.

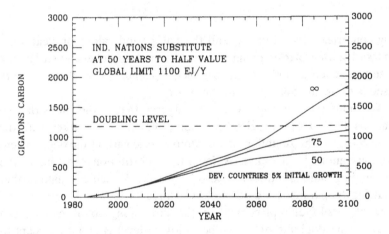

Figure 7.3. Calculated, cumulative, global emissions of carbon dioxide 1985–2100, for 5% initial annual growth of energy use and various rates of substitution in the developing countries.

process were confined to the former only. Accordingly we show in Figures 7.2 and 7.3 results of such calculations in which the replacement of the fossil fuels by nonfossil systems in the industrial countries was taken to proceed with a half-period of 50 years (half-period is the time for the replacement to reach 50%). Various rates of substitution in the develop-

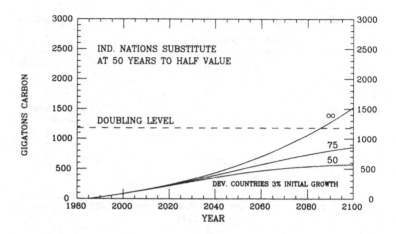

Figure 7.4. Calculated, cumulative, global emissions of carbon dioxide, 1985–2100, for 3% initial annual growth of energy use and various rates of substitution in the developing countries.

ing countries were assumed, and the half-period values (in years) are marked on the corresponding curves. The limiting global annual fuel use rate was taken as 1500 EJ (high). (See Appendix B for the procedure used in these and following calculations.)

We note immediately from these figures that if the industrialized nations alone adopt a fuel substitution policy (curves marked with the infinity sign), the doubling of the atmospheric carbon dioxide concentration cannot be avoided. A fossil-fuel substitution program in the developing nations is therefore also necessary. We can calculate further that for the parameters used in plotting Figures 7.2 and 7.3 the slowest rate of fossil-fuel replacement in the developing countries that can prevent the doubling of the carbon dioxide level is about 62 years to half substitution.

These calculations are sensitive to the choice of the assumed initial rate of growth of energy use. Thus, repeating the calculations of Figure 7.3, but with the lower initial growth rate of energy use in the developing countries of 3%, now requires a substitution time constant not longer than 81 years (see Figure 7.4). As expected, the results are also sensitive to any delay in the initiation of the fuel substitution process. A delay will entail a faster rate of substitution later on.

The exact values of the parameters chosen in the preceding examples

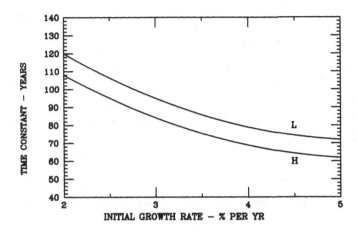

Figure 7.5. Relationship between initial energy-use growth rate and the limiting substitution rate.

are not particularly important for the present discussion. Clearly for any chosen value of the initial energy-use growth rate, we can calculate a minimum fuel substitution rate that will just prevent a doubling of the carbon dioxide level. The calculated relationship between these is shown in Figure 7.5. The main point is the demonstration that within a certain range of magnitudes (the area below the curves of Figure 7.5), there is an infinite set of combinations of energy-use growth rates and fuel substitution rates, which if implemented can avoid a doubling of the atmospheric carbon dioxide level.

At this point, it is worthwhile considering whether it is at all feasible to achieve the required values of the substitution rates. To obtain an appreciation of the magnitude of the task of introducing the new technologies at the required rate, we present in Figures 7.6 and 7.7 the calculated required annual capacity of non-fossil-energy sources and their rate of addition for each group of nations. The curves plotted in Figures 7.6 and 7.7 were calculated using the same parameters used in Figures 7.2–7.4.

The quantities in Figures 7.6 and 7.7 may be supplied on a permanent basis only by a combination of nuclear and renewable sources (Table 7.1). The mix of the two may theoretically range from an extreme maximum nuclear fraction to a zero-nuclear all-renewable system. In reality some mixture of the two, which will vary from country to country, will

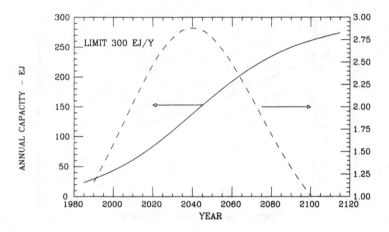

Figure 7.6. Non-fossil-energy capacity required by the industrial nations (time constant for substitution = 50 yr ; right scale = annual additions, EJ).

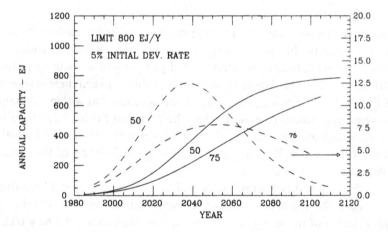

Figure 7.7. Non-fossil-energy capacity required by developing nations (time constants for substitution indicated against the curves: solid curves – left scale; dashed curves – right scale, which gives annual additions in EJ).

be found to be optimal, but we shall not proceed here with the further disaggregation of the two major groups to individual countries.

To illustrate the magnitude of the challenge of introducing the amounts of non-fossil-energy sources shown in Figures 7.6 and 7.7, we quote the following examples. To provide 1 EJ equivalent of additional primary energy per year from a nuclear source (see discussion in Section

1.3 regarding the method of expressing nuclear energy in terms of its equivalence to fossil fuels) would require the commissioning of about seventeen 1000-MW(e) reactors. To provide 1 EJ (100 TWh of electricity) per year from a photovoltaic source would require the installation of about 520 km² of cells.

We note from Figure 7.6 that the maximal rate of addition of new nonfossil capacity in the industrial countries will be about 3.0 EJ/yr and will occur about 50 years from now. If all this capacity were to be supplied by nuclear power, about 50 1000-MW(e) new reactors per year would then have to be commissioned in these countries, and about 350 would have to be under construction at any given time (not counting the replacement of aging reactors). This is about 2.2 times the number that were under construction in the world in 1985, and seems not an impossible target to achieve by the middle of the next century. In reality, only a fraction of this number would be required because it is unlikely that a total nuclear energy base would be adopted.

The situation of the developing nations (Figure 7.7) is, however, much more difficult. We note the high values of the minimal annual additions of non-fossil-energy sources needed. It is very likely that such rates of deployment of nonfossil energy may be difficult to achieve.

In considering the available non-fossil-energy resources, we note that with the present technologies most of them (e.g., nuclear, wind, hydro) are only suitable for the generation of electricity. It follows that the degree of electrification of a country becomes an important factor in determining the range of energy sources available to it and the ease of introduction of the nonfossil sources.

The industrialized countries are, in general, highly electrified and possess extensive grid systems for the distribution of power. Therefore the introduction into the electricity generation sector of those nonfossil sources for which technologies already exist can make a major and immediate contribution to the desired transition away from the fossil fuels. Furthermore, the degree of electrification in these countries can be increased readily on the basis of the existing infrastructure, and in this way the contribution of the non-fossil-energy sources can be made even greater. The new developments in the field of high-temperature superconductivity may have an important impact on the further growth of the electricity sector.

Of the renewable and inexhaustible sources listed in Table 7.1 the four largest are nuclear fission, fusion, solar energy, and wind. Of these, nuclear fission is probably the nearest to large-scale application. Although not perfect yet, the present technology is usable, economically acceptable, and being deployed in the industrialized countries as an interim measure while waiting for the next, improved, generation of reactors. The present generation of reactors, with their relatively low burnup, cannot be regarded as providing an inexhaustible source of energy (see Chapter 4), although for the period under discussion there are ample fuel resources. The next generation of reactors will hopefully be not only inherently safe but also of a very high burnup or be breeders, and this source of energy will then become practically permanent. Objections to nuclear power as practiced today have been discussed fully in Chapter 4 and will not be repeated here.

Of the other sources, fusion is at too early a phase of development to permit an estimate of when it may start contributing significantly to the transition process. Most likely it will come too late to participate in it.

Solar energy, the largest of the renewable resources, is also still in the development stage, but some significant application could be possible early in the next century. If for one reason or another nuclear fission cannot be deployed, fusion and solar energy are the only backup options in sight today.

The remaining source, wind energy, is practical today and in suitable locations may start contributing its share. It is not likely, however, that it can make a major impact in most of the industrial nations at present (see Chapter 5).

So far we have discussed only the electrical sector of the energy field. Even if electrification is pushed to the limit possible with present technologies, there will still remain a large proportion of the total final energy demand in nonelectric forms (probably close to one-half) that will have to be satisfied. At the present time this demand in the transportation, industrial, and domestic sectors is satisfied by fossil-derived fuels, mostly oil and gas. This situation may be changed in one of two ways. The degree of electrification may be pushed well beyond the limits at present considered practical, or it may be changed through the development of new ways by which the nonfossil sources may satisfy the demands of the nonelectric sectors (see Chapter 6). There are no major research programs directed to these ends, and therefore it is likely that the provision of energy to the nonelectric sectors of the economy may turn out to be a major technical factor delaying the transition to the nonfossil era.

In general, the industrialized countries have the human and financial resources and the technological basis to carry out the development of the processes needed for the transition to a non-fossil-fuel energy system. Ironically, however, the key to the future fate of the fossil resources and the environmental effects of their use, does not rest with the industrial nations but depends on the evolution of the developing nations. This conclusion derives from demographic considerations and from the basic assumption that the developing countries will continue to develop and ultimately reach the present level of the European countries.

Unfortunately, the solutions available to the industrial nations and the research carried out in them are not always applicable to the developing nations, for whom the optimal mix of primary nonfossil sources will, initially at least, be most likely very different from that of the industrial nations.

As a group, the developing countries are not highly electrified and do not possess elaborate energy distribution networks. Therefore the mere installation of additional electricity generating capacity is not a sufficient solution. This fact alone greatly limits the choice of primary energy sources, and the technologies for exploiting them, open to the developing countries. A good example of this is the case of nuclear energy and fusion. Research and development in the industrialized nations, particularly in the nuclear field, has concentrated on very large units, in the gigawatt range. It is likely that fusion reactors will also be economical only in large sizes. Although these are eminently suitable for feeding the large power distribution networks of the industrialized nations, their applicability to the developing nations is limited. We note that of the 157 reactors under construction in the world at the end of 1985, only 11 were in Asia, 6 in Latin America, and 2 in Africa and the Middle East. Even these 19 reactors were being constructed with considerable help from the industrial nations.

There are other reasons why the nuclear option is not particularly attractive to most developing nations (setting aside considerations of pride and prestige). Nuclear energy as practiced today requires a very extensive technological and organizational infrastructure to ensure its safe and efficient operation. The troubles that even the most advanced countries experienced with power reactors is a good reminder of this point. Such infrastructure is at present missing in the developing countries. It can be developed, but at considerable cost and at the expense of other national needs and to their detriment.

Even if the continuing deployment of the present generation of nuclear reactors by the industrial nations is considered an acceptable interim measure, it is a moot point whether it is also a sensible policy for the developing nations. The industrial nations already possess the know-how and infrastructure for the construction and operation of the present generation of reactors (that represent an aging technology), and so can afford the continued slow deployment of additional units without loading unduly their capacity for research and development of new technologies.

The limited economic and human resources of the developing nations will probably preclude most of them from embarking in the near future on major independent nuclear development and construction programs aimed specifically at their needs and conditions. It would certainly be a waste of effort for them to start developing and acquiring the know-how of a technology that is already old and on its way out and, in any case, not particularly suited to their needs. At this point it may be well to recall that the independent development of the next generation of reactors is a task even the most advanced industrial countries have not yet tackled. It may therefore be more prudent for the developing nations to abandon the immediate nuclear option and to decide to wait for the next generation of reactors that will be developed. Although in general it is likely that the developments will be directed to the needs of the industrial nations carrying out the work, some features may be relevant also to the developing nations. For example, a modular inherently safe reactor will require a simplified infrastructure for its operation, may better fit the smaller networks of developing nations, and be easier to adapt to their needs.

If we turn away now from the consideration of nuclear power, we find that other energy sources are far less dependent on large national power distribution networks. Examining Table 7.1 again, we note the wide choice of renewable sources, some with features that are particularly attractive to the developing group of nations. We have mentioned before (see Chapter 5) that most of the untapped hydroelectric potential is located in these countries. The technologies for exploiting this source are known and are certainly within the capabilities of most nations. Although some of the hydroelectric schemes are huge undertakings and require an extensive distribution network to handle the large amount of electricity generated, there is a considerable potential of smaller schemes. Units in the range of power below 1 MW can, and are, being installed in many locations, and their construction, operation, and maintenance are within the capabilities of the local governments. The total installed capacity of units less than 1 MW in 1984 was reported as over 11,100 MW (WEC 1986).

Wind power is also widespread in its availability, and the technology has been improved recently to the point where large wind farms can be constructed. This source is particularly suitable for small networks, although the problems arising from the intermittency of the wind still need solving. The distributed nature of this resource is also of particular advantage to countries lacking large national power grids.

The potential of biomass in its various forms is greater in many developing countries than it is in the industrial ones, and the technologies for their exploitation should not present a hurdle. Here again the distributed nature of the production and consumption of this resource presents no disadvantage to the developing nations.

The immense size of the task of introducing even these relatively simple processes can be illustrated by a further example. To add 100 TWh per year of wind-generated electricity (equivalent to 1 EJ of fossil-fuel energy) would require the annual installation of about half a million 120-kW windmills (assuming 20% capacity factor).

Although the challenge of developing and deploying these and other technologies is formidable, it can probably be met. We saw earlier in this chapter that under certain assumptions regarding the rates of energy development in both the industrialized and the developing nations, as well as the implementation of a vigorous maximal substitution policy in the former, a more leisurely rate of introduction of new technologies is possible in the latter nations.

If we take, as an example only, the case of an initial growth rate of energy use in the developing countries of 3% annually and a slow nonfossil substitution rate of 81 years to half-value, we find that for the first 30 years or so it will be necessary to install an average of 1.3 EJ/yr of new non-fossil-energy capacity. If this amount is divided among the hydro, wind, solar, and biomass sources using technologies that already exist or are about to be perfected, the task does not seem impossible. The production of the components and devices needed for such a program can form the basis for new local industries, which will provide ample and continuing employment for many years. All this will take up much money, manpower, materials, and time. Whether it can all be done within the time constraints of the transition from fossil to nonfossil-energy systems depends on the degree of international cooperation and help.

We have shown in general terms that under certain conditions it is possible to effect the transition from the present system, based on fossil fuels, to a future one based on renewable and inexhaustible energy sources, without exceeding the limit set for effective fossil resource size (some 50,000 EJ). But because of the many restrictions and constraints required on a global scale, the process is unlikely to happen spontaneously. If left alone to the market forces, development will be by a series of relatively short-range projects, with many stops and starts, leading along the path of intermediate fossil-based energy systems to an inevitable doubling of the carbon dioxide level and to other energy crises.

To skip the development of the intermediate technologies, countries must adopt some long-range and global policy. In formulating such a policy one of the most difficult factors to assess is that of the atmospheric carbon dioxide and the weight that must be assigned to its environmental effects. At present, because of the insufficient understanding of the very complex issues involved in the climatic effects of atmospheric trace gases, no clear and and generally acceptable decisions can be made.

Carbon dioxide emission differs from most other global man-made environmental abuses, such as the release of radioactivity. In the latter case the effects are universally recognized as detrimental, and therefore it is possible, even if not always easy, to generate international action to control them. But the climatic effects of increased carbon dioxide concentration in the atmosphere will not affect all nations equally. In fact, it is quite possible that for some countries or regions the effects will be deemed as beneficial. Under these conditions it will be more difficult to arrive at agreed global policies, particularly because by some countries they may be perceived as involving sacrifices and a departure from locally optimal solutions. The only reason for such nations to support a global policy for the control of carbon dioxide emissions would be from a much broader view and from an appreciation of the political destabilization and upheavals that could follow a major global climatic change. Such considerations must require a degree of political sophistication and maturity that is not too common.

Decisions on such a complex and sensitive issue will only be possible if absolutely reliable and acceptable data are available. Because of this the research effort into the carbon dioxide problem must be international, with the broad participation of industrial and developing nations. Because no action to curb fossil-fuel use is likely before the results of such research become available, its urgency is obvious.

The main long-range research effort at present is directed to nuclear energy (mainly breeder reactors and fusion), with very little to renewable sources (see Table 7.3). The imbalance of this effort can be criticized on two scores. The emphasis of practically all these programs is on the generation of electricity, to the almost total neglect of the much greater nonelectric, final energy-using sector. Relatively little effort is devoted globally to the development of technologies that will substitute for the fossil fuels in the nonelectric sectors of the economy. The other point of criticism relates to the total reliance on the nuclear and fusion options. The many problems besetting nuclear power have already been alluded to many times in this book, and the future of fusion is still by no means clear.

This research and development program coupled with the absence of a clear policy on the limitation of carbon dioxide emissions makes it very likely that a considerable delay in the transition away from fossil fuels may be unavoidable, and the doubling of atmospheric carbon dioxide concentration inescapable.

Because of the difficulties foreseen in the implementation of a timely substitution plan, some studies were carried out in recent years regarding the possibility of control of carbon dioxide emissions without requiring a complete change to a non-fossil-energy system (Seidel and Keyes 1983). The choices are very limited, but even if all they achieve is a gain of some time by minimizing the interim emission of carbon dioxide they could be useful. The type of actions that could be taken to reduce carbon dioxide emissions into the atmosphere range from "technical fixes" for removing this material from flue gases and sequestering it, to changes in the pattern of fossil-fuel use, and finally to the introduction of new processes on the basis of existing know-how. Recent studies (Seidel and Keyes 1983; Steinberg et al. 1984) show that, unfortunately, no purely technical fix is feasible.

Reference to Table 7.2 shows that the rate of carbon dioxide emission varies greatly between the different fossil fuels. This opens the theoretical possibility of control by adjusting the global pattern of use of fossil fuels by regulation or taxation. However, detailed calculations (Edmonds and Reilly 1983; Seidel and Keyes 1983) show that nothing short of a total ban on coal and oil shale use will make an appreciable differ-

Table 7.3. *Research and development in energy by some[a] OECD governments, 1985 (millions of 1985 $ U.S.)*

	Canada	Germany	Italy	Japan	Netherlands	Spain	United Kingdom	United States	Total $	Total %
Conservation	50.36	14.22	19.35	12.29	19.20	34.40	30.94	173.60	409.95	6.61
Oil and gas	151.17	7.81	0.0	57.87	0.69	8.17	28.11	38.40	305.21	4.92
Coal	23.79	91.65	0.50	146.46	19.32	17.29	5.39	257.90	594.44	9.59
Nuclear nonbreeder	109.66	186.21	213.34	769.63	15.23	14.35	82.67	577.70	2,008.06	32.40
Nuclear advanced[b]	6.95	195.92	308.61	440.70	10.05	1.06	159.43	617.40	1,785.20	28.80
Solar	9.22	31.85	10.70	36.51	4.18	23.52	1.28	105.60	240.95	3.89
Other renewable	14.28	22.00	6.30	31.99	28.98	15.71	14.37	114.70	267.92	4.32
Other	11.12	9.85	10.44	60.40	17.12	34.22	35.69	371.20	586.59	9.46
Total	376.57	548.52	569.24	1,557.86	114.77	124.14	364.06	2,256.30	6,198.33	100.0

[a]Governments spending over $100 million a year.
[b]Breeders and fusion.

Source: OECD/IEA (1986).

ence to the doubling date of carbon dioxide in the atmosphere. Less severe restrictions on fuel use provide only a few years delay. These conditions are difficult to fulfil, and the international record so far for dealing with similar problems is very discouraging. We may conclude that it will be quite difficult to achieve a considerable delay of the doubling date of the carbon dioxide concentration by controlling the fossil-fuel mix.

The only possibility remaining, therefore, of slowing the rate of emission of carbon dioxide is by accelerating the rate of replacement of the fossil fuels by non-fossil-energy sources. The main impact of the shift away from fossil fuels will have, perforce, to await the long-range developments of new energy systems, but much can be done in the interim by implementing measures based on existing knowledge. Such actions must be taken by industrial and developing countries (see Figures 7.2 and 7.3).

Initially at least, the brunt of such a program will have to be borne by the industrialized countries, for at the present time they are the major energy consumers, and only they possess the necessary know-how and infrastructure for an immediate start. The major aim of such an accelerated program of substitution in the industrialized countries would be to reduce their demand for fossil fuels, particularly for oil, as fast as possible. The fossil fuels thus released will enable the developing nations to prolong their fuel substitution process and thus increases its chance of success. This is important, because in the long run it is what happens in the developing countries that will determine the future global energy demand pattern and its environmental consequences.

The most immediate impact on the transition away from fossil fuels that the industrial nations might make can come from two lines of action. The first is energy conservation, particularly in those sectors consuming oil. The second is through a maximal nuclear electricity policy, coupled with vigorous electrification schemes (see Chapter 4).

A similar large-scale deployment of nuclear power in the developing nations cannot be implemented on a short time scale and, in any case, is not optimal there (see earlier discussion). Other technologies will have to be developed that are more suitable for rapid implementation in these countries. A second aim of the industrial nations must then be the increase in the research and development effort of other, nonnuclear, technologies suitable both to their own needs as well as to those of the developing countries. We shall not repeat here the list of possibilities.

We would like, however, to draw attention again to the imbalance of the research programs shown in Table 7.3. Although, as a principle, it is important that technologies for use in the developing countries be developed by their own people, a parallel program running in the industrial countries can be of great help.

There is another important reason why the industrial nations should take the lead in the fuel substitution activity. Any policy of accelerated substitution will probably deviate from an optimal short-term plan based on purely economic considerations. The difference between the two represents the price society must pay to ensure its long-term aims. It is inconceivable that the developing nations will agree to adopt a policy of energy source substitution that on a short-term basis may appear to them to be less than optimal unless they are convinced of its long-term value and if the sharing of the burden by the industrial nations has been demonstrated.

We presented in the last few paragraphs a possible course of action for the transition from the present energy system to a future one. It does require the diversion of considerable material and human resources, and it will not happen without some long-range planning. It does not require universal international agreements, for which the chances are slim, but rather the cooperation of the major energy-consuming nations and those mainly involved in the research and development efforts. If these research programs are successful, it will not be necessary to legislate internationally to enforce fuel substitution. The economic forces could then come into play, but it is essential that a selection of technologies be available for them to play with.

The point was already made that the success of any such transition plan depends on the leadership and wisdom of the major industrial powers. If they abdicate this role, two crises may overtake humanity. The first may be the disruptions caused by the climatic changes that could follow the increased carbon dioxide load in the atmosphere. The second may arise from the delay in the development of the technologies needed for the orderly transition to the nonfossil systems. The upheavals that will occur when the fossil fuels become scarce and no alternative energy sources are ready to take their place will pose a grave threat to civilization as we know it.

The alternative to such global policies is not a return to the "good old days" when everything was small and beautiful, but a move to large-scale misery for the bulk of the world's inhabitants. It is significant that the calls for a return to a primitive world are invariably expressed in the most sophisticated, well-fed, and well-heeled nations. We do not hear them from those people who still live under these supposedly idyllic conditions. They should be the most happy people.

Additional reading

Eden, R. J. et al (1981). *Energy Economics.* Cambridge University Press, Cambridge.

Edmonds, J. and Reilly, J. M. (1985). *Global Energy,* Oxford University Press, New York.

Grenon, M. (1981). *The Nuclear Apple and the Solar Orange.* Pergamon Press, New York.

Häfele, W. (1981). *Energy in a Finite World.* Ballinger, Cambridge. Mass.

MacCracken, M. C. and Luther, F., eds. (1985). Projecting the Climatic Effects of Increasing Carbon Dioxide. U.S. Department of Energy Report DOE/ER-0237. Washington, D.C.

National Research Council (1983). *Report of the Carbon Dioxide Assessment Committee.* National Academy Press, Washington, D.C.

Seidel, S. and Keyes, D. (1983). Can We Delay a Greenhouse Warming? U.S. Environmental Protection Agency, Washington, D.C.

Trabalka, J. R. and Reichle, D. E., eds. (1986). *The Changing Carbon Cycle, A Global Analysis.* Springer-Verlag, New York.

Appendixes

Appendix A: Energy conversion factors
Table of multipliers and energy conversion factors

A. Multipliers

Factor	Prefix	Symbol
10^3	kilo	k
10^6	mega	M
10^9	giga	G
10^{12}	tera	T
10^{15}	peta	P
10^{18}	exa	E

B. Conversion factors (to joules)

1 BTU	=	1055 J
1 kcal	=	4184 J
1 kWh	=	3.6 MJ
1 MeV	=	1.602×10^{-13} J
1 MWD	=	86.4 GJ
1 TWyr	=	31.54 EJ
1 TOE	=	44.76 GJ
1 TCE	=	29.29 GJ
1 m³ nat. gas	=	37.26 MJ

C. Conversion factors (from EJ)

1 EJ	=	947.9×10^{12} BTU
	=	239×10^{12} kcal
	=	31.71 GWyr
	=	11.57×10^6 MWD
	=	31.71×10^{-3} TWyr
	=	22.34×10^6 TOE
	=	34.14×10^6 TCE
	=	26.84×10^9 m³ gas

Appendix B: Procedure for calculating cumulative energy use

In Chapters 6 and 7 we present results of calculations of cumulative energy use under various assumptions. These calculations are relatively straightforward and are within the means of anyone possessing an electronic hand-held calculator. For the benefit of readers who might be tempted to carry out similar calculations based on their own models or assumptions, we present below an outline of the procedure used.

Recall (Sections 6.1 and 6.3) that the basic assumption is that energy use and the substitution among energy forms follow logistic behavior, that is they will obey the equations

$$f = \frac{e^{(\alpha t - \beta)}}{1 + e^{(\alpha t - \beta)}} \tag{1a}$$

or

$$\ln \frac{f}{1-f} = \alpha t - \beta , \tag{1b}$$

where α and β are constants to be determined by the model and assumptions. The constant β is determined from the value of f at time f_0:

$$-\beta = \ln \frac{f_0}{1 - f_0} .$$

The value f_0, in turn, is determined from the known annual energy use q_0 at time zero and the assumed limiting value L (see Section 6.1), thus:

$$f_0 = \frac{q_0}{L} .$$

To determine α, we need one other value of the annual energy use, q_t at time t. If a certain rate of initial annual increase in the energy use r is assumed to be valid over a period τ, then we have

$$q_t = q_0 r^\tau .$$

This value of q_t is then used to calculate f_t and, hence, by equation (1b), the value of α,

$$\alpha = \frac{1}{\tau} \left[\ln \frac{f_t}{1 - f_t} + \beta \right] .$$

Knowing α and β, we can determine the cumulative energy use for any time by using the integrated form of equation (1a):

$$Q_t = \frac{L}{\alpha} \ln \frac{e^\beta + e^{\alpha t}}{e\beta + 1}.$$

This is suitable for calculating overall cumulative energy use as well as cumulative use of a particular energy resource, provided it remains constant through the interval considered. All that is necessary in the latter case is to multiply the value Q_t by that fraction.

Bibliography

Aldrich, M. J., Laughlin, A. W., and Gambil, D. T. (1981). Geothermal Resource Base of the World : A Revision of the EPRI Estimate. Los Alamos Scientific Laboratories, Univ. of California.

Annual Energy Review (1986). Energy Information Administration, Washington, D.C.

D'Arsonval, A. (1891). Utilisation des forces naturelles. Avenir de l'électricité. *Rev. Sci.* **17**. 370.

Avron, M. (1984). The outdoor cultivation of the halotolerant algae *Dunalinella:* A model of biosolar energy system for useful chemical products. In C. Sybesma, ed., *Advances in Photosynthesis Research. Preceedings 6th International Congress on Photosynthesis (Brussels)*, Vol. 2. pp. 745–53, W. Junk, The Hague.

Bennett, L. L. (1985). Economic Performance of Nuclear Plants: How Competitive? *IAEA Bull.* **27**: 40.

Bockris, J. O'M. (1980). *Energy Options.* Wiley, New York.

Bolin, B. (1979). Global Ecology and Man. In *Proceedings of the World Climate Conference*, p. 24. World Meteorological Organisation, Geneva.

Broecker, W.S., Takahashi, T., and Tsung-Hung Peng (1985). Reconstruction of Past Atmospheric CO_2 Contents from the Chemistry of the Contemporary Ocean: An Evaluation. Report DOE/OR-857. U.S. Department of Energy, Washington, D.C.

Bronicki, L.Y. (1984). Twenty Five Years Experience with Organic Working Fluids in Turbomachinery. *Proceedings of the International VDI Seminar on New Working Fluids for Energy Engineering, Zurich. VDI Ber.,* **539**: 685–96.

Burger, J. (1973). L'exploitation des pyroschisteson schistes bitumeux données générales et perspectives d'avenir. *Rev. Inst. Français Pétrole* **28**: 315–72.

Carr-Saunders, A.M. (1936). *World Population: Past Growth and Present Trends.* Clarendon Press, Oxford.

Cataldi, R. and Sommaruga, C. (1986). Background, present state and future prospects of geothermal development. *Geothermics* **15** (v): 3.

Chant, V.G. (1981). Two Global Scenarios: Evolution of Energy Use and the Economy to 2030. Report RR-81-35, International Institute for Advanced System Analysis, Laxenburg, Austria.

Clark, W.C., ed. (1982). *Carbon Dioxide Review: 1982.* Oxford University Press, New York.

Claude, G. (1930). Power from tropical seas. *Mech. Eng.* **52:** 1032.

Cohen, R. (1980). Energy from thermal gradients, *Oceanus* **22:** 12.

Colombo, U. and Bernardini, O. (1979). A Low Energy Growth Scenario and the Perspective for Western Europe. Report prepared for the commission of European communities panel on low energy growth.

Dickinson, R.E. and Cicerone, R.J. (1986). Future global warming from atmospheric trace gases. *Nature* **319:** 109.

DiPippo, R. (1985). In *Geothermal Electric Power. The State of the World – 1985. 1985 International Symposium on Geothermal Energy,* Geothermal Research Council, Davis, CA.

DOE (1982). Projected Costs of Electricity from Nuclear and Coal-Fired Power Plants. DOE/EIA-03561, Energy Information Administration, Washington, D.C.

Edmonds, J. and Reilly, J. (1983a). Global energy production and use for the year 2050. *Energy* **8:** 419–32.

(1983b). A long-term global energy-economic model of CO_2 release from fossil fuel use. *Energy Econ.* **5(2):** 74–88.

(1983c). Global energy and CO_2 in the year 2050. *Energy J.* **4:** 21.

(1985). *Global Energy – Assessing the Future.* Oxford University Press, New York.

Edmonds, J.A., Reilly, J., Trabalka, J.R., and Reichle, D.E. (1984). An Analysis of Possible Future Atmospheric Retention of Fossil Fuel Carbon Dioxide. U.S. Department of Energy Report DOE TR-013, National Technical Information Series, Springfield, VA.

EPRI (1985a). *A State-of-the-Art Study of Nonconvective Solar Ponds for Power Generation.* EPRI Report AP-3842, Palo Alto, CA.

EPRI (1985b). Heber: Key to Moderate-Temperature Geothermal Reserves. *EPRI Journal,* January/February.

Fisher, J.C. and Pry, R.H. (1970). A Simple Substitution Model of Technological Change. Report 70-C-215; Technical Information Series, General Electric Company, Schenectady, NY.

Friedlander, G., Kennedy, J.W., Macias, E.S., and Miller, J.M. (1981). *Nuclear and Radiochemistry, 3d ed.* Wiley, New York.

Frisch, J.-R., ed. (1983). In *Energy 2000–2020: World Prospects and Regional Stresses.* World Energy Conference, Graham & Trotman, London.

Glaser, P. (1969). Satellite solar power stations. *Sol. Energy* **12**: 353.

Gold, T. (1985). The origin of natural gas and petroleum, and the prognosis for future supplies. *Ann. Rev. Energy* **10**: 53–77.

Gold, T. and Soter, S. (1980). The deep-earth gas hypothesis. *Sci. Am.* **242**: 154–61.

(1982). Abiogenic methane and the origin of petroleum. *Energy Explor. and Exploit,* **1** (2): 89–104.

Goldemberg, J., Johansson, T.R., Reddy, A.K.N., and Williams, R.H. (1985). An end-use oriented global energy policy. *Ann. Rev. Energy* **10**: 613–88.

Gustavson, M.R. (1979). Limits to windpower utilization. *Science* **204**: 13–17.

Häfele, W. (1981). *Energy in a Finite World – A Global System Analysis.* Ballinger, Cambridge, MA

Häfele, W., Holdren, J.P., Kessler, G., and Kulcinski, G.L. (1977). Fusion and Fast Breeder Reactors. Report RR-77-8, International Institute for Advanced System Analysis, Laxenburg, Austria.

Halbouty, M.T. (1983). Reserves of Natural Gas Outside the Communist Bloc Countries. In *Proceedings 11th World Petroleum Congress,* vol. 2, pp. 281–92. Wiley, New York.

Harth, R.E. and Boltendahl, U. (1981). The chemical heat pipe: EVA and ADAM. *Interdisciplinary Sci. Rev.* **6(3)**: 221–28.

International Atomic Energy Agency (1979). *Status and Prospects of Thermal Breeders and Their Effect on Fuel Utilisation.* Technical Reports Series, No. 195.

(1980a). Fuel and Heavy Water Availability. Report of the INFCE Working Group 1, Vol. 1.

(1980b). Enrichment Availability. Report of INFCE Working Group 2, Vol. 2. Reprocessing and Plutonium Handling. Report of INFCE Working Group 4, Vol. 4.

(1980d) Fast Breeders. Report of INFCE Working Group 5, Vol. 5.

(1980e). Spent Fuel Management. Report of INFCE Working Group 6, Vol.6.

(1980f). Waste Management and Disposal. Report of INFCE Working Group 7, Vol.7.

(1980g). Advanced Fuel Cycle and Reactor Concepts. Report of INFCE Working Group 8, Vol.8.

(1982a). *Guidebook on the Introduction of Nuclear Power.* Technical Reports Series, No.217.

(1982b). *Data Series #1,* Vienna.

(1986). *Data Series #1,* Vienna.

International Energy Agency (1978). *Steam Coal – Prospects to 2000.* OECD, Paris.

(1986). *Coal Information 1986,* OECD/IEA, Paris.

Isaacs, J.D. and Schmitt, W.R. (1980). Ocean energy: forms and prospects. *Science* **207**: 265–73.

Jaffe, D., Friedlander, S., and Kerney, D. (1987). The LUZ Solar Electric Generating Systems in California. In *Proceedings of the ISES World Conference, Hamburg.*

Keeling, C.D., Bacastow, R.B., and Wharf, T.P. (1972). Measurement of the Concentration of CO_2 at Mauna Loa. In W.C. Clark (ed.), *Carbon Dioxide Review: 1982*, pp. 377–85. Oxford University Press, New York.

Keyfitz, N. et al. (1983). Global Population and Labour Force (1975 – 2075) Report ORAU/IEA-83-6(M), Institute of Energy Analysis, Oak Ridge Associated Universities.

Kreith, F. and Bharathan, D. (1988). Heat Transfer Research for Ocean Thermal Energy Conversion. *J. Heat Transfer,* **110**(5).

Kreith, F. and Kreider, J (1978). *Principles of Solar Engineering.* McGraw-Hill, New York.

Kreith, F. and Meyer, R.T. (1984). Solar Thermal Power Towers. *Int. J. Solar Energy,* **2**: 385–404.

Kuuskraa, V.A., Hammershaimb, E.C., and Stosur, G. (1983). The Efficiency of Enhanced Oil Recovery Techniques: A Review of Significant Field Tests. In *Proceedings 11th World Petroleum Congress.* Vol. 3, pp. 387–411, Wiley, New York.

Lennard, D.E. (1984). *OTEC – Progress and Prospects.* Energy Options Conference, London.

Liss, P.S. and Crane, A.J.(1983). *Man Made Carbon Dioxide and Climatic Change. A Review of Scientific Problems.* Geo Books, Norwich, England.

Lynette, R. (1985). Wind Power Stations: 1984 Performance and Reliability. EPRI Report AP-4199, Palo Alto, CA.

MacCracken, M.C. and Luther, F.M., eds. (1985a). Detecting the Climatic Effects of Increasing Carbon Dioxide. Report DOE/ER-0235, United States Department of Energy.

   (1985b). Projecting the Climatic Effects of Increasing Carbon Dioxide. Report DOE/ER-0237, United States Department of Energy.

Manne, A.S. and Schrattenholzer, L. (1985). International energy workshop: a summary of the 1983 poll responses. *Energy Journal* **5**(1): 45–64.

Marchetti, C. and Nakicenovic, N. (1979). The Dynamics of Energy Systems and the Logistic Substitution Model. Report RR-79-13, International Institute for Advanced System Analysis, Laxenburg, Austria.

Marland, G. (1982). The Impact of Synthetic Fuels on Global Carbon dioxide emissions. In W.C. Clark (ed.), *Carbon Dioxide Review: 1982.* Oxford University Press, New York.

Masters, C.D., Root, D.H., and Dietzman, W.D. (1983). Distribution and Quantitative Assessment of World Crude Oil Reserves and Resources.

U.S. Geological Survey Open File Report 83-728; also in *Proceedings of 11th World Petroleum Congress*, Vol. 2, pp. 220–39, Wiley, New York.

Matveev, A.K., ed. (1975). Oil shales outside the Soviet Union. In *Deposits of Fossil Fuels*, Vol. 4. G.K. Hall, Boston, Mass.

Meyer, R.F. et al., eds. (1984). *The Future of Heavy Crude and Tar Sands.* McGraw-Hill, New York.

Meyerhoff, A. (1979). Proved and Ultimate Reserves of Natural Gas and Gas Liquids in the World. Preprint PD12, paper #5; 10th World Petroleum Congress, 1979

Montagna, A.A. (1978). In *Future Sources of Organic Materials – CHEMRAWN 1*, p. 21, St-Pierre, L.E., and Brown, G.R., eds., Pergamon, 1980.

Nordhaus, W.D., and Yohe, G.W. (1983). *Future Paths of Energy and Carbon Dioxide Emissions, in Changing Climate.* Report of the Carbon Dioxide Assessment Committee, National Academy Press, Washington D.C.

Nuclear Energy Agency/International Atomic Energy Agency (1982). *Uranium. Resources, Production and Demand.* OECD, Paris.

(1983). *Uranium. Resources, Production and Demand.* OECD, Paris.

OECD (1980). *Energy Balances 1975–1980.* OECD, Paris.

OECD/IEA (1982). *Report by the Coal Advisory Board.* OECD, Paris.

(1986) *Energy Policies and Programmes of IEA Countries: 1985 Review.* OECD, Paris.

OGJ Report (1983). *Oil and Gas Journal*, 26 December 1983.

Perry, A.M. (1979). World Uranium Resources. WP-79-64, International Institute for Applied System Analysis, Laxenburg, Austria.

Peterka, V. (1977). Macrodynamics of Technological Change: Market Penetration by New Technologies. Report RR-77-22, International Institute for Advanced System Analysis, Laxenburg, Austria.

Pry, R.H. (1973). Forecasting the Diffusion of Technology. Report 73-CRD-220, General Electric Company, New York.

Putnam, P.C. (1953). *Energy in the Future.* Van Nostrand, Princeton, NJ.

Rose, D.J., Miller, N.M., and Agnew, C. (1983). Global Energy Futures and Carbon Dioxide Induced Climatic Changes. Report MITEL 83-015, Massachusets Institute of Technology, Cambridge, MA.

Seidel, S. and Keyes, D. (1983). 'Can We Delay a Greenhouse Warming?' Office of Policy Analysis, Office of Policy and Resources Management, Washington, D.C.

Steinberg, M. et al. (1984). A Systems Study for the Removal, Recovery and Disposal of Carbon Dioxide from Fossil Power Plants in the U.S. Report (DOE/CH/0016-2), U.S. Department of Energy, Washington, D.C.

Swift-Hook, D.T. and Taylor, R.H., eds. (1983). *IEEE Proc.* **A130**:9. Special issue on windpower.

Thomson, W. (Lord Kelvin) (1851). *Math. and Phys. Papers*, Vol.1, p. 179.

Trabalka, J.R. ed. (1985). Atmospheric Carbon Dioxide and the Global Carbon Cycle. Report DOE/ER-0239, U.S. Department of Energy, Washington, D.C.

Trabalka, J.R. and Reichle, D.E., eds. (1986). *The Changing Carbon Cycle, A Global Analysis.* Springer-Verlag, New York.

U.S. Army Corps of Engineers (1981). *National Hydroelectric Power Resources Study,* Vol. 12. Washington, D.C.

U.S. Congress (1978). Enhanced Oil Recovery Potential in The United States. O.T.A. Report OTA-E-59, Washington, D.C.

Weinberg, A. M. et al. (1985). *The Second Nuclear Era: A New Start for Nuclear Power.* Institute for Energy Analysis, Oak Ridge, Tennessee. Praeger, New York.

Wick, G.L. and Schmitt, W.R. (1977). Prospects for renewable energy from the sea. *MTS Journal* **11;** 16–21.

Wilcox, W.F., ed. (1931). *International Migrations, Vol. 2.,* National Bureau of Economic Research, Washington, D.C.

(1940). *Studies in American Demography.* Cornell University Press, Ithaca, N.Y.

World Bank (1984). *Energy Statistics.* Washington, D.C.

World Energy Conference (1983). *Oil Substitution: World Outlook to 2020.* Graham & Trotman, London, and Oxford University Press, Oxford.

(1986). *Survey of Energy Resources.* Holywell Press, Oxford.

Zweibel, K. (1986). Photovoltaic cells. *Chem. Eng. News* **67:** 34–48.

Index

Printed in the United States
By Bookmasters